TRAGEDY IN MOUSE UTOPIA

The sorcerer lurks within

AN ECOLOGICAL
COMMENTARY ON HUMAN UTOPIA

by

JOHN R. VALLENTYNE, PH. D.

TRAFFORD
PUBLISHING™

Note for Librarians: A cataloguing record for this book is available from Library and Archives Canada at www.collectionscanada.ca/amicus/index-e.html
ISBN 1-4120-5633-0

Printed on paper with minimum 30% recycled fibre. Trafford's print shop runs on "green energy" from solar, wind and other environmentally-friendly power sources.

TRAFFORD
PUBLISHING™

Offices in Canada, USA, Ireland and UK
This book was published *on-demand* in cooperation with Trafford Publishing. On-demand publishing is a unique process and service of making a book available for retail sale to the public taking advantage of on-demand manufacturing and Internet marketing. On-demand publishing includes promotions, retail sales, manufacturing, order fulfilment, accounting and collecting royalties on behalf of the author.

Book sales for North America and international:
Trafford Publishing, 6E–2333 Government St.,
Victoria, BC V8T 4P4 CANADA
phone 250 383 6864 (toll-free 1 888 232 4444)
fax 250 383 6804; email to orders@trafford.com
Book sales in Europe:
Trafford Publishing (UK) Limited, 9 Park End Street, 2nd Floor
Oxford, UK OXI IHH UNITED KINGDOM
phone 44 (0)1865 722 113 (local rate 0845 230 9601)
facsimile 44 (0)1865 722 868; info.uk@trafford.com
Order online at:
trafford.com/ 05-0531

10 9 8

ACKNOWLEDGMENTS

I THANK EDITH SMITH, my philosopher son Peter, and Shannon Gale for editorial guidance in preparing this book; my physiatrist son Stephen for the story of Mr.J. K.; my daughter, Jane, for alerting me to emerging problems of human health; and Togwell Jackson, geochemist-artist, for the frontispiece. John and Ann Mahan of Grand Rapids, Michigan, Tom Muir of Burlington, Ontario, and Sue Korol of Victoria, British Columbia, made many useful suggestions that I followed. My daughter, Anne Marie, and my wife, Ann, did the final editing. Garrett Hardin, David Suzuki and Jan Barica prodded me to write the book.

For permission to quote from the following published works I am indebted to:

- Oxford University Press for *A Sand County Almanac* (1949) by Aldo Leopold, and for *The Art of War* by Sun Tzu translated and edited by Samuel B. Griffith (1963);

- Lilly Fabilly Osborne and Harper/Collins Publishers for *The Temper of our Time* (1969) by Eric Hoffer;

- Random House Inc. for *This Endangered Planet* (1972) by Richard A. Falk;

- W. W. Norton & Co. for *Civilization and its Discontents* (1961) by Sigmund Freud, translated by James Stacey;

- the Journal of the Royal Society of Medicine for the article by John B.Calhoun (1973): *Death Squared:the explosive growth and demise of a mouse population*. Proc.Roy.Soc. Med. 66: 80-88; and

- Harvard University Press for permission to quote from: THE WORKS OF WILLIAM JAMES: ESSAYS IN RADICAL EMPIRICISM, Fredrick Burckhardt, General Editor and Fredson Bowers, Textual Editor, Copyright© 1976 by the President and Fellows of Harvard College.

- I owe a great debt to the many persons not cited in the text who contributed so much to the development of ecology and evolution in the past two centuries.

CONTENTS

THE AUTHOR AS SCIENTIST

〜 JOHN R.VALLENTYNE COMPLETED AN un-
dergraduate degree in biology in 1949 at Queen's
University in Kingston, Ontario, and a Ph. D. at
Yale with G. Evelyn Hutchinson in 1953. He taught
biology and ecology at Queen's from 1952 to 1958,
and ecology and evolution at Cornell from 1958 to
1966. He was Senior Scientist at freshwater research
centers in Winnipeg, Manitoba, and Burlington,
Ontario, from 1966 to 1992, and President of the
International Association of Limnology from 1974
to 1980. In 1972 he convinced the governments of
the United States and Canada to get phosphates out
of detergents, and in 1978 to adopt a "man-in" ecosystem approach to managing environmental quality. His awards include a Guggenheim Fellowship
in 1964-65; the Rachel Carson Award of The Society
of Environmental Toxicology and Chemistry in
1992 "given to an individual who has substantially
increased public awareness to an extent that has re-
sulted in the redefinition of policies and practices";
and a Lifetime Achievement Award of the American
Society of Limnology and Oceanography in 2002.

He has authored over 100 scientific publications including a 1974 book, *The Algal Bowl: Lakes and Man*, now available in expanded form with David W. Schindler as senior author as: *The Algal Bowl: Phosphorus and the Control of Eutrophication in Lakes.*

Visit Johnny's web site at http://www. johnnybiosphere.ca to learn about Johnny's extensive travels, outrageous stories, and provocative talks in schools (1988 photo).

THE AUTHOR AS PERSON

Jack, or JR, as he is known to friends and family, was born in Toronto in 1926 and grew up from the age of nine in Brantford, Ontario. On week-ends and in summers it was canoeing, swimming, and fishing with boyhood friends. He joined the Canadian Army at 17 and was quickly identified as a sharp-shooter, but World War II ended before he was old enough to fight. He married Ann Tracy in 1947, and is the father of five children, two adopted. As an amateur actor he played Ferdinand in *The Tempest*, the lieutenant in *The Caine Mutiny Court Martial*, and the Latin lover in *Barefoot in the Park*. He became an avid downhill skier at 38, a melanoma survivor at 50, and a planetary activist as Johnny Biosphere at 54. He retired in 1992 and in 1996 received an Honorary D. Sc. degree from McMaster University. He lives happily with his extended family in Hamilton, Ontario, enjoying a glass (or two) of red wine with his evening meal.

Humpty Dumpty sat on a wall,
Humpty Dumpty had a great fall.
All the king's horses and all the king's men,
Couldn't put Humpty together again.

PREFACE

THE FERMI PARADOX

THE STRONG ANTHROPIC COSMOLOGICAL PRIN-
CIPLE (Breuer 1991) asserts that "The laws of nature
applying to our universe must be such that life
can arise under them and endure for long periods
of time, for if it were otherwise we would not be
here."

Enrico Fermi, Italian-born American Nobel
Laureate in Physics, could have been thinking along
these lines when he was experimenting under the
football stadium of the University of Chicago in
World War II on how to build a nuclear bomb. He
pondered: *If, as many scientists believe, intelligent life
has evolved at many times and places in the Universe,
why have we not received any communications to that
effect?*

One of the many possible explanations of the
Fermi Paradox is that technological civilizations
may be commonly created in the Universe, yet be so
widely separated in space and time as to preclude

the possibility of ever communicating even at the speed of light.

If planetary civilizations are randomly distributed in the Universe, the lack of communication could imply that the average lifetime of planetary technological civilizations may be less than a few thousand years. For that to be the case *without exception* it follows that technological civilizations must contain the seeds of their own destruction.

Technological civilizations have existed in Egypt and China for more than four thousand years, and a premeditated Search for Extraterrestrial Intelligence (SETI) has been in operation since 1959 without any positive findings. Of course, absence of evidence should never be taken as evidence of absence; nevertheless, if the short lifetime hypothesis is valid, it implies that our global technological civilization could be nearing an end.

Some potential seeds of destruction have already erupted into impending threats. These threats include: the runaway global growth of human technology and population, deceit on all levels from one's self to others, personal greed coupled with global indifference, and nuclear war. Any one of these, or others, could bring an end to human aspirations to play a significant role in the evolution of the Universe.

This book claims that the fulfillment of human short-term desires has fueled a runaway growth cycle in which technology and population have spurred each other to ecologically unsustainable heights. If the pleasure-seeking drives that have spurred the runaway cycle on its destructive course are viewed collectively and endowed with a personality, what emerges is the idea of a Sorcerer within, with us as the Sorcerer's apprentices. If this suicidal behavior continues, Human Utopia, like Mouse Utopia, will self-destruct.

Can an ecologist challenge the economic doctrine of global economic growth without limit and win? Consider the biblical story of David and Goliath (1 Samuel: 17), but with David as the symbol of nature and Goliath as the symbol of the runaway global growth of human technology and population.

Goliath is decked out in heavy armor from head to foot. David rejects armor, saying that he has not proved its worth. Instead, David drops five smooth stones from a brook into his shepherd's bag. Running toward the giant, sling in hand, David drives one of the stones deep into Goliath's forehead. Goliath falls to the ground. David takes Goliath's sword and cuts of his head.

This book is about the life or death of a global technological civilization. Read it! Read it! And read it again! Then, act in accordance with your new understanding and perceptions.

This above all,–to thine own self be true;
And it must follow, as the night the day,
Thou canst not then be false to any man.

Hamlet, Act I, Scene IV,
Polonius to his son, Laertes.

PART I

THE TRAGEDY

⁓ OVER A PERIOD OF four and a half years eight mice in an enclosure designed as a model of human technological civilization increased to a population of 2200 and then declined to extinction. Stresses arising from overcrowded conditions in Mouse Utopia barred young mice from access to social roles in the colony. As a result, young mice grew up without knowing how to behave as mice. The behavioral stresses in Mouse Utopia resemble those arising in Human Utopia from the runaway growth of technology and population. Failure to meet the psychological needs of the young at critical stages in development has been common to both situations. Does this mean that we are on a path leading to self-detruction?

1. TRAGEDY IN MOUSE UTOPIA

If you seek the welfare of your children, let
them experience a little cold and hunger.
A Chinese Proverb

⌒ ON JULY 9, 1968, John B. Calhoun (1917-1995),
mammalian ecologist and Chief of the Section on
Behavioral Systems at the National Institute of
Mental Health in Bethesda, Maryland, started an
experiment as a model of global technological civilization. Calhoun introduced eight 48-day-old, Balb
C albino house mice, four males and four females,
into an enclosed space that, taking certain literary
liberties, we can call Mouse Utopia.

Calhoun provided the mice with all the resources they needed to live – an unlimited supply
of uncontaminated air, food and water, abundant
bedding material, and comfortable housing. In response, over a period of four and a half years the

mice rapidly increased to a population of 2,200 and then declined, slowly and inexorably, to extinction.

That is a very strange result. Provide mice with everything that mice need to live and the colony dies out? It doesn't make sense.

There was no evidence of predators or disease, two factors that commonly limit the growth of populations. Even more strangely, almost half the living space remained unused at the height of the population explosion. The history of Mouse Utopia is a mystery. The purpose of this chapter is to state the facts, and take a first cut at unravelling the mystery.

The eight mice used to seed Mouse Utopia came from a breeding colony at the U.S. National Institutes of Health, where extreme precautions had been taken to exclude micro-organisms such as *Salmonella* that can cause epidemics in mouse populations. Mouse Utopia was thus protected from devastating diseases – as are we – by a superb, technologically driven health protection system. Because of inbreeding, all mice in Mouse Utopia were created more-or-less equal.

Mouse Utopia covered a floor area of about nine square meters. This provided just enough room for Calhoun (who was short and thin) to bend over, turn around, and gingerly step in one or another direction. The area was enclosed by four walls, each

1.37 meters high. There was no ceiling. Calhoun entered and exited from above.

The walls were lined with sixteen evenly spaced, four-story apartment buildings. These were provided with convenient walk-up stairs leading at each level to four cozy, one-room apartments, each capable of comfortably housing fifteen mice. Food hoppers and water bottles, each yielding nourishment on demand, were located half-way up each apartment building. The central area provided a place for social interactions.

The floor between the apartment buildings was covered with ground corn cobs to soak up urine and paper strips for bedding material. Calhoun himself provided a clean-up service every six to eight weeks for removal of excrement, dead animals and soiled bedding. While this would be considered grossly inadequate by modern human standards, it was, according to Calhoun, quite acceptable for mice.

It is easy to imagine how vast and wondrous this brave new world must have seemed to the eight mice introduced into Mouse Utopia. There was comfortable and conveniently located housing for 4,000 mice; protection from predators, disease, and inclement weather; a garbage removal service (Calhoun); an undertaker (Calhoun); an unlimited supply of bedding material, food, water and fresh, climatically controlled air; and places to mate with members of the opposite sex and socialize with neighbors.

Each mouse was marked for individual identification. Over four and a half years Calhoun methodically recorded individual and social behavior, occupied and unoccupied living quarters, births, deaths, age distribution, fights, love-affairs, harassment, murder, and all the other events that demographers like to keep track of in human populations. Autopsies were performed on dead mice. Mice in parallel colonies were sacrificed to monitor internal changes accompanying growth of the experimental population.

The history of Mouse Utopia fell naturally into four phases shown in Figure 1.1. These were identified by Calhoun in exemplary scientific fashion as: A (days1 to 104), B (days 105 to 314), C (days 315 to 560), and D (days 561 to 1610). As a guide to what follows it is useful to keep in mind that 10 days in the life of a mouse corresponds to about a year in the life of a human.

Phase A covered the 104 days preceding birth of the first litter. This was a time of great social turmoil until the eight immigrants became adjusted to each other and to their new technological surroundings.

Phase B began with birth of the first litter. After that, the population increased exponentially, doubling every 55 days. The increase was still exponential in *Phase C.* but the population doubling time slowed to 145 days. *Phase D* began with Mouse Utopia's peak population of 2200 mice.

By the start of *Phase D*, few females in the colony had ever conceived and all had lost the ability to rear young beyond the day of birth. The fate of Mouse Utopia was sealed by day 560. The last surviving male died on day 1610, and the last surviving female on day 1644, four and a half years after the original settlement of the colony.

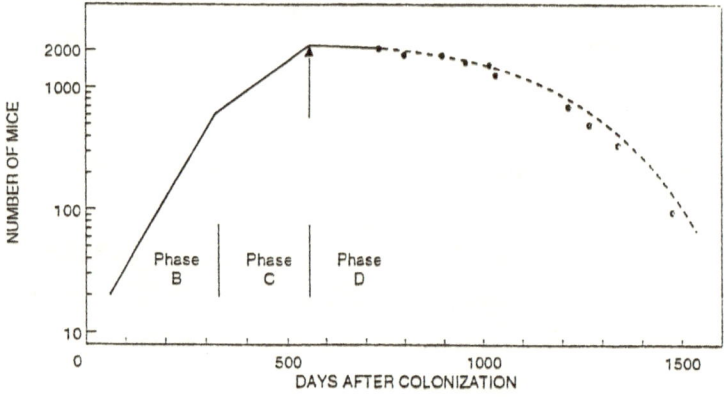

Figure 1.1. History of a population of mice in a closed Utopian Universe from Calhoun (1973). The broken line represents Calhoun's estimate of future numbers made on day 700 on the basis of observed mortality to that time. Observed points after day 1000 are lower than earlier projected due to removal of 150 mice for other studies.

Calhoun interpreted what happened as follows: Toward the end of *Phase B* (days 105 to 314), many dominant males, exhausted from having to defend their territories from hordes of young males looking for mates, lost interest in defending their territories. With no one on guard, nursing mothers aggressively protected their families. By the end of *Phase B* there were three times as many young mice aspiring to enter social groups as there were vacancies in the socially established older groups.

During *Phase C* the incidence of conception in females declined and the resorption of fetuses increased. Maternal behavior was disrupted. The result was increasing numbers of miscarriages. Some mothers in desperate searches for quieter quarters abandoned young that fell on the way.

Prematurely rejected, first by their fathers, then by their mothers, and then by established groups in the community, the young grew up without knowing how to behave, personally or socially, as mice. Disruptions appeared in the social behavior of the young toward the end of *Phase B*, long before the mouse population peaked at the end of *Phase C*. Some mice expressed their frustration in aggression. Others, apparently discouraged by lack of access to social roles in life, withdrew from mouse society.

In *Phase C* withdrawn males increasingly aggregated in "pools" in the central area where they lounged around indifferent to territorial defense and sexuality. Every now and then one would venture out for food or drink; otherwise they remained listless unless disturbed by nearby outbursts of pointless violence. Despite vicious attacks the withdrawn males made little effort to flee even when severely wounded. Later, many of the attacked mice became attackers.

By the middle of *Phase C* all young had been prematurely rejected by their mothers. Some withdrawn males hung around the food hoppers where they sat alone, watching. Another group of males, "the beautiful ones", lived together in out-of-the-way apartments, leading a kind of autistic life without participating in sex or competing for territory. Their coats were sleek and well-groomed. Wounds were few or absent.

Many young females, denied access to meaningful social roles in Mouse Utopia, also withdrew from mouse society. Some became unusually aggressive. Others spent their days hiding in vacant upper-level apartments less preferred by females with litters where they exhibited similar forms of listless, indifferent behavior to the "beautiful one" males.

Withdrawn mice had nothing to do and nowhere to go. The future, if they thought about it at all, must have seemed dull and unappealing. Calhoun viewed them as arrested juveniles. Lacking social bonding, they were physically adult, but immature behaviorally and neurologically.

Even at the height of the population explosion, the vacancy rate of apartments was twenty percent. Availability of physical living space was thus not an important factor in the demise of Mouse Utopia. Yet, in Calhoun's experienced eyes, the colony had been assigned to the death-row before the end of *Phase C*. Extinction was only a matter of time.

What went wrong in Mouse Utopia? What caused the little colony to self-destruct? Are there lessons here for humanity?

2. WHAT WENT WRONG IN
MOUSE UTOPIA?

Nature, to be commanded, must be obeyed.

Francis Bacon, *Novum Organum*

As a result of the increasingly crowded conditions in Mouse Utopia, mice rubbed shoulders with other mice everywhere they went. Body contact was further increased by a pronounced tendency for the mice to occupy preferred locations and to aggregate at food hoppers where other mice were already present. According to Calhoun, the mice developed a new definition of the feeding situation, one that included the presence of other mice.

This "pathological togetherness" with its consequent changes in the rhythm of eating, interfered with the sequential development of biologically vital behaviors, such as courtship, mating, parenting and aggression. As a result, social behavioral sequences that were inappropriate or incomplete became fixed at an early stage of development. The survival instincts of the mice had been dulled. The tragic outcome was that the mice looked like mice, yet did not know how to behave as mice in ways that were personally satisfying and advantageous to survival of their species.

Alert to the psychological and social implications of his findings, Calhoun did not hesitate to speculate on the relevance of events in Mouse Utopia to modern trends among children. These trends included declining trust and lack of empathy for others, lack of bonding with society, lack of respect for their society's value system, lack of opportunity for social development, and inability to control aggression.

Calhoun even dressed up some of his scientific publications with biblical quotations. For example, in presenting an account of events in Mouse Utopia to the Royal Society of Medicine in London, England, he began with the four horses of the Apocalypse in *Revelation* and ended with a re-arranged quotation from *Proverbs 3: 13, 17, and 18:* "Happy is the man who finds wisdom and the man who gains understanding. Wisdom is a tree of life to those who lay hold of her. All her paths lead to peace."

Calhoun set 1984 as the human date "beyond which the opportunity for decision making and designing to avoid population catastrophe might be quickly lost." *Note that 1984 was not the predicted year of human apocalypse, but the last year for decisions to avert apocalypse.*

What are we to make of this? Was Calhoun as naive as some biologists believe in extrapolating from rodents to humans? Did he go off the deep end by putting human and religious twists on the results of his mouse experiment?

I spent an afternoon with Calhoun in his laboratory in 1975. I found him to be a quiet, thoughtful and reflective person. I do not think he was naive. I think he knew exactly what he was doing, and probably how the experiment would end before it began.

Calhoun wanted to warn humanity of the disastrous effects of overcrowding on human behavior. He had studied the ecology of wild mice and rats long before Mouse Utopia began. To Calhoun, Mouse Utopia was a demonstration more than an experiment.

Calhoun believed that we grossly underestimate the importance of social behavior in "lower" animals. He considered that social rank among rodents in a colony is just as significant in shaping social behavior as it is for humans. From this it followed that disruptions of normal behavioral states in both mice and humans can result in unhealthy states of mind and body.

Ian McHarg, American city planner, compared human environments of health and pathology in Philadelphia in his insightful book *Design with Nature* (1969). McHarg found a positive correlation between "social diseases" (homicide, suicide, drug addiction, alcoholism, robbery, rape, unaggravated assault, juvenile delinquency, and infant mortality) with high population density, poverty, and related factors.

McHarg's approach was complementary to Calhoun's. Calhoun extrapolated from mice to humans, whereas McHarg extrapolated from humans to muskrats and rats based on studies made by Dr. Jack Christian, ecologist at the Philadelphia zoo. Christian related his findings to increases in cardiovascular and renal disturbances induced by stress under crowded conditions.

McHarg quoted Dr. P. Leyhausen, a European ethologist who spent nearly five years in overcrowded prisoner-of-war camps in Europe during World War II. Leyhausen concluded that: "the basic forces of social interaction and organization are *in principle* identical and there is a true homology between Man and Animal throughout the whole range of vertebrates."

Behavioral parallels between human and mouse populations cannot be dismissed with facile assertions that we are not mice or that we are smarter than mice. The important question is the extent to which mice and humans share behavioral characteristics that make both species susceptible to collapse when provided with easy access to unlimited resources.

On the other hand, biologists will be quick to note that Calhoun's mice came from a highly inbred stock, reducing the genetic variety available for selection during the experiment. Biologists are also wary of inter-species comparisons. Mice may be mammals with behavioral patterns similar to those of other vertebrates, but they are also four-footed, small-sized, small-brained creatures that are adapted to living under conditions completely different from those of humans.

But wait!

Perhaps we should not be too quick in coming to a final judgment based on human perceptions. Our arrogance in asserting superiority over other kin on the tree of life may not be warranted. Rationality may not be as powerful as many think relative to unconscious, instinctive, and habitual influences on our behavior.

Most population biologists interpret Calhoun's experiment as a demonstration of the disastrous effects of overcrowding. To them, the results seem so obvious that the experiment was hardly worth doing. To be sure, the experiment was both these things; but it was also much more than that.

It was a demonstration of how technology and

population can interact, uncontrollably, to the detriment of individuals and ultimately to extinction of the whole colony. The message is clear: provide an easy life for a population, and the population will self-destruct.

The history of Mouse Utopia shows that misplaced adaptiveness is not a uniquely human trait. Mice, too, must have internal values that are reflected in their behavior. Many of these values may, as in our case, lie hidden in the unconscious recesses of their minds. When mice are prevented from translating those values into appropriate forms of behavior, the result is psychological derangement. The same may be true for Human Utopia which, in this book, refers to the course of development of humanity, mainly based on western civilization, since the fourteenth century.

Population biologists have generally failed to note that Calhoun was part of the experiment. Like an overly benevolent technological god, Calhoun provided all that was needed to make life easy for mice. Then the mice, programmed by their gene-based evolutionary history, did the rest in their new technological surroundings. In the end, there was nothing left for them to do but to age until called back into the Earth by the clammy hand of death.

Calhoun concluded his account of Mouse Utopia with these words:

> For an animal so simple as a mouse, the most complex behaviors involve the interrelated set of

courtship, maternal care, territorial defense and hierarchical intragroup and intergroup social organization. When behaviors related to these functions fail to mature, there is no development of social organization and no reproduction. As in the case of my study, all members of the population will age and eventually die. The species will die out. For an animal so complex as man, there is no logical reason why a comparable sequence of events should not also lead to species extinction. If opportunities for role fulfilment fall short of the demand of those capable of filling the roles, and having expectancies to do so, only violence and disruption of social organization can follow. Individuals born under these circumstances will be so out of touch with reality as to be incapable even of alienation. Their most complex behaviors will become fragmented. Acquisition, creation and utilization of ideas appropriate for life in a post-industrial, cultural-conceptual-technological society will have been blocked. Just as biological generativity in the mouse involves this species' most complex behaviors, so does ideational generativity for man. Loss of these respective complex behaviors means death of the species.

Viewed from a higher plane, a meta-analysis of the strange sequence of events in Mouse Utopia suggests that the greatest threats to Human Utopia could arise from factors that:

- unleash reproductive potential by removing constraints to growth

- are nonmaterial, difficult to quantify, and damaging to mind and spirit

- have effects that are delayed by one or more generations, and

- once set in motion are practically impossible to reverse.

Many of these conditions now exist in urban populations. By trying to command nature without obeying her rules, it would seem that we, like Calhoun's mice, have been caught in a trap. Of course, it can be objected that Calhoun fueled the population explosion of mice by creating, technologically, a regime of unlimited resources; whereas in our case, having created the technology, we think that we can control it.

But can we? Will we? In time?

3. A GLOBAL VILLAGE?

If the sailors become too numerous the ship
sinks.

An Arabic Proverb

⌒ CALHOUN DESIGNED MOUSE UTOPIA as a model of human population growth in a technologically
contrived environment. Yet, he also knew that models are abstractions of reality. Models simplify some
things and exaggerate others to show features that
are not readily apparent in all the complexities of
life.

For example, every mouse in Mouse Utopia
could interact personally with every other mouse;
whereas, even with air travel and email, this is not
the case for humanity in the Biosphere. Similarly,
Mouse Utopia was small, even for a mouse, and
uniform; whereas the Biosphere is immense beyond
human belief and diverse.

Nevertheless, some similarities are common to Mouse Utopia and overpopulated metropolitan areas such as Tokyo, Sao Paulo, Mumbai (Bombay), New York City, Los Angeles, London, Beijing and Mexico City. When scaled up from a 100 gram mouse to a 70 kilogram technological human being the density of people in some areas of these and other mega-cities is about the same as that of mice in Mouse Utopia.

To take Mouse Utopia seriously as a model of our possible future, as Calhoun obviously did, it is necessary to think like a mouse and to feel what it would have been like to live as a mouse in Mouse Utopia. The first intellectual baggage to jettison in preparation for this mental voyage is our insistence that we in our world are vastly superior to mice in their world.

We are not.

The symbol of the evolutionary tree, now amply documented by DNA, suggests that all living organisms can trace their ancestry back to the first reproductively successful forms of life on Earth. Thus, in terms of adaptation, survival and reproductive continuity, it cannot be said that any one species is vastly superior to another – except relative to species that are extinct.

Now, let us suppose that Sophocles had been able to draw on the results of Calhoun's experiment as background for a play. Perhaps he would have written the play in four acts corresponding to the four phases identified by Calhoun as A, B, C, and D. Maybe he would have called them *The Age of Wonder* (days 0-104), *The Age of Innocence* (days 105-314), *The Age of Despair* (days 315-560), and *The Age of Doom* (days 561-1610).

Imagine now, that you are one of those small, blind, hairless, happy-go-lucky mice born early in *The Age of Innocence* in Mouse Utopia. After being squeezed out of the warm placental swimming pool that was home for the first three weeks of life, you would, for no apparent reason, gulp a breath of air and head straight for one of your mother's several milk-laden breasts.

From that day on you would be drenched in all the warmth and tenderness of a mouse's motherly love until the time of weaning. Your father would protect the family from intruders. As you grew up, you would have friends to play with, new acquaintances to make, and mates of the opposite sex to engage in courting behavior.

Life would be full of fun, excitement and deep, uninterrupted sleep. There would be no debilitating sicknesses or unnatural deaths in the family, no cats to terrify and torment you, no shortages of fresh uncontaminated air, food or water. Even by the best of human standards, life would seem pretty good.

Next, suppose that you survived to become a parent, grandparent, or even great grandparent in the early part of *The Age of Despair* (days 315-560). Can you imagine the agony of watching the rising incidence of pillage, harassment and brutality on the streets? Kids saucily sticking their tongues out at you in the town square or even biting your tail when their mothers weren't looking?

What would you think of sons who became drop-out fathers, daughters who as mothers battered their kids, or grandsons who went around town indiscriminately mounting males and females? Just think of it: the other day while you were out grabbing a bite to eat at a fast food restaurant, a sex-crazed adolescent nosed his way into your apartment and raped your baby daughter!

As a pregnant grandmother, wouldn't you be depressed by the rising numbers of miscarriages and resorptions of fetuses in your womb? Wouldn't you begin to wonder what life was all about? In our view there was no right or wrong in Mouse Utopia because mice can't discriminate between moral and immoral behavior. But is our view of mouse behavior correct?

Consider as a human what it would be like to be constantly harassed as a mother by gangs of abrasive young hooligans with no respect for a lady. Little wonder that you would become increasingly irritable and aggressive, and that a lot of that aggression would spill over to your children.

Fed up with life in your apartment wouldn't

you want to take off for a more respectable neighborhood, hoping against hope to avoid harassment along the way?

Finally, imagine that you were a mouse born in the latter part of *The Age of Despair.* Life in your mother's womb would have been strained and stressful. Four brothers and sisters would lie dead or dying beside you.

Your father would have deserted the family long before you were born. No one could blame him for leaving. What is a dominant male supposed to do when he can't control the hordes of lecherous teenagers breaking into his home and trying to mount his wife and children?

If you grew up in such an appalling environment, it would only be to enter a world that had no place for you. As a young female, rejected and fearful of life, you might cringe in a dark corner of a deserted fourth floor apartment. As a young male, you might listlessly hang around in the town square with other rejected males, or just sit idly watching the passing parade of mice in a fast food restaurant.

As a "mouse psychologist" Calhoun wanted to show how overcrowding leads to rejection and loss of role fulfilment in adolescents. He wanted to show how this loss results in inability to perform the complex and socially vital behaviors of courtship, territorial defense and maternal care that are requisite to survival of vertebrate species. He wanted to show that it was not physical space, but social interactions that determined the carrying capacity of Mouse Utopia for mice – and, by analogy, may determine the carrying capacities of mega-cities or the Biosphere for humans. He wanted to forestall the fates, but his prophecies, like those of Cassandra, were destined to be mocked.

4. FROM MICE TO MEN

All nature is but art, unknown to thee;
All chance, direction, which thou canst not see;
All discord, harmony not understood;
All partial evil, universal good.

Alexander Pope, *An Essay on Man*

CALHOUN PROVIDED THE TECHNOLOGY for the design, construction and maintenance of Mouse Utopia. Every time a mouse was added to the population, the technological system automatically furnished the necessary food, water, bedding, shelter, and all the rest. The experiment was, by its very nature, a mixture of technology and biology.

Perhaps you can now see where all this is leading. If so, don't run away. It could be important for the Biosphere and for the future evolution of Human Utopia.

The objective, now, is to identify crucial times

for human policy decisions on the growth of technology and population in Human Utopia based on the findings in Mouse Utopia. Remember that ten days in the life of a mouse correspond to about a year in the life of a human being.

Significantly, by the end of *The Age of Innocence* (days 105-314) there were three times more young mice aspiring to meaningful social roles than there were vacancies in Mouse Utopia, and the ratio was increasing at an alarming rate. *In effect, the fate of the little mouse colony was sealed during the Age of Innocence.*

Based on a mouse-day to human-year comparison, this would imply that policy decisions on the control of growth of human population and technology would have to be made during the first twenty-one years of life in mega-cities or technological civilizations in the Biosphere.

The result of this calculation is obviously absurd, and the reasons are not hard to find. Advanced technology came all at once in Mouse Utopia, whereas it took centuries to develop in the Biosphere. Also, a female mouse can deliver six to ten pups in a litter and have five or six litters between puberty and menopause. This is a much higher reproductive potential than that of a human female a century ago when families with thirteen children were not uncommon – and can still be found in some parts of the world today.

It is clear that the mouse-day to human-year conversion is useless as a guide in relating events in Mouse Utopia to the need for policy decisions in the Biosphere.

But wait!

Let us not discard the comparison too quickly. Two hundred and ten days in *The Age of Innocence* in Mouse Utopia could easily correspond to well over a century in human time if three additional factors are taken into account: the instantaneous introduction of technology in Mouse Utopia, the greater reproductive potential of mice, and the many measures adopted by human societies to reduce the worst impacts of overpopulation.

The doubling times of populations could provide a more realistic basis for comparing events in Mouse Utopia to events in the Global Village because generation times are called into account. The sequence of population doublings in Mouse Utopia from the beginning to end of *The Age of Innocence* was 20, 40, 80, 160, and 320 days – five doublings in all.

The question now is: What time-spans does this correspond to for mega-cities and for humanity as a whole? The best way to calculate these connections could be by working backwards from the end of *The Age of Innocence* for humanity through five "halfings" of human population.

Most people would probably accept that *The*

Age of Innocence in Human Utopia ended for mega-cities on 9-11, 2001, if not before. Globally, it was on August 6, 1945, with the explosion of the first atomic bomb used in war.

First, let us first examine the case for mega-cities. For a mega-city with a population, say, of 20 million people in 2001, five halfings of population would successively take us back to 10 million, 5 million, 2.5 million, 1.25 million, and then to 600,000 people. Significantly, there were only fourteen cities with a population of over one million people in 1900, whereas there were over four hundred in 2001!

But a city of 600,000 people is still grossly oversized for a species that evolved in tribes of about one hundred and fifty individuals and in villages of a few thousand. Evidently, the comparison is suggestive, but not exact.

Now to the planetary situation. Starting with the end of *The Age of Innocence* in 1945, five halfings of the human population of the Earth would successively take us from 2.4 billion (the world population in 1945) down to 1.2 billion, 600 million, 300 million, 150 million, and eventually to 75 million people.

The next step is to fix the global population number of 75 million to a date. This is pure guesswork at best; hard estimates of global human populations before 1900 are not available. Nevertheless, indirect estimates by some population experts suggest that the human population of the Earth may have reached 75 million some time around the thirteenth or fourteenth century.

Setting the end of *The Age of Innocence* for humanity at any time during the twentieth century and working backwards through five halfings of population would take us back to a suggested starting date for *The Age of Innocence* some time between the twelfth and fifteenth centuries. These suggested starting dates for the explosion of human technology and population roughly coincide with the appearance of bell towers in European cities (fourteenth century), Johannes Gutenberg's invention of printing based on moveable type (in the first half of the fifteenth century), and the re-discovery of America by Columbus in 1492.

Even more significantly, it was around the fourteenth century that global human population began to show the first signs of runaway growth – despite deaths from the pandemic diseases that then prevailed.

Now we seem to be getting somewhere!

Of course, these correlations could be coincidental rather than causal. Nevertheless, there may be something significant at work here. In this business it is unwise to rule anything out. You never know where a new piece in the jig-saw puzzle may lead. My aim is to get reasonably close to a complete picture. If I can't, I want to lay a foundation on which others can build.

The explosive growth of technology and human population came in a blur after the seventeenth century. Two Scotsmen prepared the way for this: James Watt (1736-1819) and Adam Smith (1723-1790).

In the nineteenth century came a whirl of railways, steamboats, photographs, telegraphs, telephones, internal combustion engines and electric lights.

In the twentieth century it was airplanes, tanks, and an explosion of cars, television sets, antibiotics, guided missiles, jet planes, computers, Internet, transistors, nuclear bombs, email, cell phones and space exploration. Many of the innovations in the twentieth century were born in World War I, World War II and the Cold War.

Wars spur inventions.

It is hard to avoid the conclusion that these regressions in the twentieth century were symptomatic of a world in *The Age of Despair*. The behavior of entire nations was berserk, and people immune to propaganda knew it.

For those who doubt that we have been in an era comparable to *The Age of Despair* in Mouse Utopia for most or all of the twentieth century, consider the following trends in Ontario in the 1980s and 1990s. In perusing these trends, note that

Canada for several years in the 1990s was at the top of the United Nations' list of preferred countries to live in. These trends were:

- increasing breakdown of families since 1950

- more than 40% of children living apart from at least one of their genetic parents

- one-fifth of all children in Ontario with at least one psychological disorder, and two-thirds of these with two or more psychological disorders

- a rapidly rising incidence of reported child abuse, street gangs, and juvenile murderers

- a general loss of confidence in governments, and

- a lack of meaningful role-fulfilment through employment opportunities for young people.

Doesn't this tabulation resemble Calhoun's description of life in Mouse Utopia in the early stages of *The Age of Despair*?

The most significant causal features common to the two utopias are: the provision of unlimited resources, the geographically enclosed environment, the breakdown of families, technology, the bounded nature of the systems, and separation from nature. Much of the rest seems to follow directly from these – including the stressful life of parenting and developmental crippling so cogently described in the human case by Paul Shepard in *Nature and Madness* (1983) and in the domestication of our species by John Livingston in *Rogue Primate* (1994).

In fact, the situation in Human Utopia would be far worse than it is, had it not been for innumerable human measures, unavailable to mice, to ease the situation. These include: birth control, health care, child care, protection from child abuse, adoption, imprisonment for criminal behavior, welfare, employment insurance, job training, environmental protection, police protection, legal systems, and others.

One can even say that these attempts to reduce problems caused by runaway growth in Human Utopia were major causes of runaway growth. Ironically, the day of reckoning was not avoided; its most devastating effects were merely postponed.

Alexander Pope described our limited knowledge well in *An Essay on Man*. There is a lot more understanding built into our bodies than we can ever comprehend. Sometimes, in respect to complex situations such as those described in this book, the intuitive (right) sides of our brains provide better insights into what is happening than the logical (left) sides of our brains.

Some right-brain images that come to mind in signaling the beginning, middle and end of *The Age of Innocence* in Human Utopia are: the legendary Johannes Faust who lived in the first half of the sixteenth century; the publication of Mary Wollstonecraft Shelley's novel *Frankenstein* in 1818; and the sinking of the *Titanic* in 1912.

If any of these or other events or conjectures seem too early or late, feel free to choose others on your own. Just one word of caution: beware of unrealistic hopes based on unbridled optimism, indifference or denial. Hope springs eternal because the dead cannot speak.

If you find it depressing to think that we may have been caught in a socio-behavioral trap of our own making for practically all of the twentieth century, do not sweep the bad news under the carpet, hunt scapegoats, or kill the messenger.

There are grounds for hope, but before examining them it is essential to evaluate the extent to which aberrant forms of social organization have undermined the health and integrity of our species and the Biosphere.

5. MOTHERLY LOVE

We see with our eyes.
We know with our hearts.

Fraggle Rock

⟫ PERHAPS YOU MAY BE dismayed, astonished, and even insulted to find yourself compared to a mouse. If so, this chapter may help you to accept that, despite differences in form, vertebrates have many behavioral similarities. The discussion of *marasmus*, largely drawn from Ashley Montagu's book *Touching* (1971) is a good place to start. Marasmus is a medical term derived from a Greek word meaning "wasting away". The term "foundlings" refers to abandoned children of unknown parents.

Before 1920, virtually one hundred percent (*yes, one hundred percent!*) of infants under the age of two years who were admitted to foundling institutions in the eastern United States, and kept there for any appreciable time, died from marasmus.

The cause of marasmus remained obscure until Dr. Fritz Talbot of Boston visited The Children's Clinic in Dusseldorf a few weeks before the start of World War I. Talbot noticed a plump elderly woman carrying a puny infant on her hip. In response to his queery, the director replied, "Oh, that is old Anna. When we have done everything we can medically for a baby, and it is still not doing well, we turn it over to Old Anna and she is always successful."

In 1915, Dr. Henry Dwight Chapin, a noted New York pediatrician, documented the incredible rates of infant deaths in foundling institutions in ten American cities. He confirmed the cause as a lack of tender, loving care. Subsequently, over decades, the incidence of marasmus in children's hospitals declined. Dr. James L Halliday, a leading American psychiatrist, described the desperate needs of infants to be cuddled, stroked and talked to in *Psychosocial Medicine: A Study of the Sick Society* (1948).

In 1958 Harry Harlow, then President of the American Psychological Association, presented a lecture to the Association on *The Nature of Love*. Strangely, he noted, most psychologists had bypassed love as a primary mammalian drive along with hunger, thirst, elimination, and sex. Harlow exposed this gross error in experiments on young rhesus monkeys in association with surrogate, artificially contrived, mothers. He showed that body contact through "stroking" and "object-clinging" early in life was vital to normal development. In fact, body contact was even more important than whether nursing was from a mother's breast or feeding from a surrogate (technological) mother dressed in terry cloth and warmed by a small internal light bulb. He used photographs to show the importance of skin touching skin to the young of other species – including elephants, hippopotamuses, crocadiles and snakes. Others have confirmed and extended Harlow's findings of the importance of touch in early life.

Mary Carlson, a neural scientist at the Harvard Medical School, found that monkeys raised by natural mothers that they could see, smell, and hear – *but not touch* – developed what she described as "an autistic-like syndrome with grooming, self-clasping, social withdrawal and rocking."

Recently, Dr. Michael Meaney, professor of medicine at McGill University and Associate Director of Research at the Douglas Hospital Research Center in Montreal, extended Harlow's approach to studies on rat behavior. Rat pups were separated into two groups based on the frequency of licking and grooming by their mothers. Those with less body contact were more susceptible to stress and stress-related illnesses later in life. Meaney also discovered that stroking with a small paint brush could substitute for licking by a mother.

In a parallel human study Meaney found that people who remembered having poor relationships with their mothers when young had high levels of stress hormones and stress-related illnesses throughout life. In both rats and humans the debilitating effects were traced to stress hormones produced in the brain.

The similarity of these findings in monkeys, rats and humans raises an interesting question. Is it possible that a lack of "motherly love" could have caused the tragic history of Mouse Utopia to unfold? It is beginning to look as though the answer could be yes.

Erik Erikson (1902-1994) is often considered to be the first true life-span developmental psychologist. He identified eight stages between birth and death at which behavioral crises emerge. In each stage he believed that the *ego* plays a crucial role in resolving the crises. The role of the *ego*, Erikson claimed, is to mediate the struggle between inner biological urges and outer societal constraints. If the *ego* fails to resolve these crises at any stage, future development is impaired and the mentally crippled individuals pay for it for the rest of their lives.

Do mice have *egos*? Do they have developmental stages comparable to those in humans with crises that have to be resolved before proceeding from one to the next?

The answer in both cases seems to be *yes*; but as far as I know, no one besides Calhoun has made a life-span psychological analysis of the behavior of mice.

Has Calhoun's "demonstration-experiment" of Mouse Utopia ever been repeated? The answer is: Yes. It is being repeated right now in Human Utopia, as Calhoun accurately predicted more than 30 years ago. The cause of social disruptions common to both Mouse Utopia and Human Utopia is a runaway cycle of technology and population, *jointly.*

For *ecological stresses* induced by the runaway cycle in Human Utopia see chapters 11, 12 and 16 in this book. For a detailed account of *social stresses* of the runaway cycle in Human Utopia (although Wooding does not use either of these terms) consult Scott Wooding's excellent book, *The Parenting Crisis* (2005).

The television industry in North America has responded positively to the parenting crisis. There are now (2006) kids' stories on Treehouse TV all day-long for children. There are also popular hour-long adult TV programs (such as Dr. Phil and Oprah Winfrey) on week-days for families caught up in problems of obesity, drug abuse addictions, poor communication, and the like.

PART II

THE BIOSPHERE

⌒ AN ECOLOGICAL AND EVOLUTIONARY framework for viewing human life on Earth is needed to resolve stresses induced by the growth of technology and population. The contrast of *environment* and *ecosystem* as *house* and *home*. The story of *The Man and the Suit* as a metaphor for life in the built environment. *Ego-System* views and ecosystem views contrasted. The notion of a personal ecosystem. Discovery of the Biosphere as an organized system. Historical development of the Biosphere compressed into a week. Ten precious legacies left by the Biosphere. Why the concept of the Biosphere is essential to the continuity of human life on Earth.

6. FROM EGO-SYSTEM TO ECOSYSTEM

> An optimist is a person who turns on a light
> that isn't there. A pessimist comes along and
> blows it out.
>
> Source Unknown

🖋 UNGRACIOUSLY PUSHED OUT OF our mothers' wombs at birth, we gulp air and begin independent lives as pure, unadulterated, self-centered *Ego-Systems.*

Loveable, yes; but completely wrapped up in ourselves. We assume that the Universe is here to serve us. We are surprised and even outraged when experience shows that it is not.

It may be the same for all mammals, perhaps even for all forms of life. How could it be otherwise when organisms are genetically programmed to start life with "me first"?

But "me first" is not "me only". Behavior based

on "me only" is like a death certificate. Evolution is based on an unconscious presumption of cooperation in biospheric interest. That is more like "me too" or "all together now".

At the other end of the planetary spectrum, *Biosphere* is the scientific name for the enormous living and life-supporting system in the outer part of the Earth. We came from it. We live within it. And we cannot live without it. It took the Biosphere more than three billion years to prepare the way for our species on Earth.

Throughout practically all of the history of our species we have operated in ignorance of these facts. We have been trapped by self-fulfilling *Ego-Systemic* desires. Many people still persist in believing that they are special creations, separate from nature. They forget or do not know how much they depend on the Biosphere for their life-support.

Hippocrates (about 460 B.C. to about 370 B.C.), the Greek father of western medicine, did not hold these views. He was convinced that everything is connected both inside and outside the body. His book *Air, Waters and Places* was as much a treatise on human ecology as on medicine.

Hippocrates considered food, occupation and climate to be influential factors in causing diseases. He had no sympathy with the idea, current in his day, that disease is a punishment sent by the gods. He believed that the role of the physician is to understand and assist nature, both within the body and between the body and its operational environment. In modern scientific terms, his outlook was circular causal, holistic, and ecosystemic.

The word *environment* is far too disconnected, too ill-defined and too remote to be as personally meaningful as Hippocrates intended. The way we use the term *environment* makes it seem to us that it is outside.

Just as sickness defines health in the body, so does environmental pollution define health in soil, water, food and air. In both cases after-the-fact treatment is often too late to avoid irreversible effects. In contrast, if you view your environment as part of your body you will safeguard it as you would your life.

You and your personal environment make up your personal ecosystem - or, if you prefer, your extended personality - which can conceptually encompass the Biosphere and Universe. The boundaries of ecosystems are flexible and can be changed according to need. One example of this was the change in context of the Great Lakes Water Quality Agreement from water to ecosystem. Another came with views of the Biosphere in photographs of the Earth from space.

Except for sewage treatment on land, the Great Lakes Water Quality Agreement of 1972 was almost exclusively focused on water. In contrast the revised Agreement of 1978 defined the Great Lakes Basin Ecosystem as "the interacting components of air, land, water and living organisms, *including humans*, within the drainage basin of the St.Lawrence River at or upstream from the point at which this river becomes the international boundary between Canada and the United States." This was needed to provide a regulatory framework that would encourage two nations, eight states and two provinces to work together, with no loopholes, toward a goal of improved water quality in the Great Lakes. In addition, by focusing on the ecosystem concept, the 1978 Agreement linked the Great Lakes politically to the rest of the Biosphere.

Environment is like a *house*, something external that you can walk away from and leave behind. In contrast, an *ecosystem* is like a *home*, something that you see yourself in even when you are not there. An ecosystem has an added spiritual dimension that makes it qualitatively different from a house. It can be a happier place because of the caring and sharing relationships of its inhabitants and the hospitality tendered to visitors.

There is a profound difference between our cultural (built) environments and our natural environments. In the following story, the suit is a metaphor for our cultural environment.

The Man and the Suit

One day, the chief of a branch office of the GDP Corporation ripped the jacket of his suit on his way to an important meeting at the head office in Boston.

"No matter," Joe assured himself. "My old tailor will fix me up with a new suit on the spot."

Sure enough, Fred had dozens of quality suits on the rack. Joe saw two that he liked. One was a conventional grey with black stripes, but it was too big. The other was a perfect fit; but it was green, and Joe had a deep-seated aversion to green. So Joe picked the grey suit.

"Give me an hour," said Fred. "Your suit will be ready when you get back."

When Joe looked in the mirror he was aghast. One sleeve hung below his finger tips. The other ended three fingers above his wrist.

In desperation, Joe asked for the green suit. However, it turned out that another customer had come in just after Joe left and walked out wearing the green suit.

Joe was standing like a mannequin in front of the mirror when Fred raised one of Joe's shoulders and lowered the other. "Look," said Fred, "a perfect fit!"

Joe was not so sure, but as time was short he went along with Fred's suggestion.

Two of Joe's old friends were lined up behind him at the registration desk.

After Joe left, one nudged the other: "Hey!
Did you see that? I wonder what ever happened to Joe!"

"Yeah," said the other, "but did you notice
the fit of his suit?"

————•————

The message of this story is that we have adjusted psychologically and physiologically to the
cultural environments that we have created, and
in the process have become ecological and evolutionary misfits. To see the pertinence of the story
just look around at all the self-created difficulties
brought on in technological civilizations by alcohol,
tobacco, poor diets, lack of exercise, trashing the
wild environment, fetal alcohol syndrome, crime,
skyscrapers, terrorism, sexual abuse of children,
wife beating, traffic injuries, want of time, want of
spirit, want of parental love, susceptibility to pan-
demics, obesity, indoctrination by propaganda, and
so on.

The whole history of technological civilizations has been to disengage from nature, to worship money, to covet land as a commodity, to rav-
age forests, to drain wetlands, and to send wastes
downwind and downstream for as long as you can
get away with it.

The political take-over of the Biosphere was extended to the oceans by the United Nations Convention on the Law of the Sea. The Convention was opened for signature on December 10, 1982 and came into force on November 16, 1994. It permitted coastal states to extend their sovereignty to a breadth of 12 nautical miles in territorial seas, and to 200 nautical mile limits designated as "exclusive economic zones".

People everywhere have tried to reduce the human impacts of the growth of technology and population without fundamentally changing their ways. If the issue is health, the first concern is for human health. If the issue is environmental pollution, the first concern is for pollution of the human environment. If the issue is declining resources, the first concern is to increase the supply of human resources such as drinking water, food, fertilizers, petroleum, and natural gas. As human technology and population increased, the health and integrity of the Biosphere declined. Short-term gain was allowed to take precedence over long-term loss.

In the late 1960s traffic police in Tokyo wore gas masks. On every continent fish kills, oil spills, and beach closures were daily news events. Sun bathers on Mediterranean beaches in the 1960s were shocked to find their feet and towels streaked with black grease from oil-soaked sand. The bodies of bald eagles and peregrine falcons in North America were so polluted with long-lived chlorinated hydro-carbon insecticides that many individuals couldn't reproduce. In some areas environmental pollutants changed the sex of fish, reptiles, birds and mammals. In 1967 the Torrey Canyon became the first of many supertankers to break apart, dumping its massive load of Bunker C oil on rocks and sea birds between the coasts of England and France. In 1969 the Cuyahoga River in Ohio, caught fire from the oily debris of technological civilization on its surface. Late in the twentieth century acid rain from the United States was killing fish in poorly buffered lakes in Canada, acid rain from England was killing forests in Sweden, and acid rain from Germany was killing forests in Poland and Czechoslovakia.

Between 1949 and 1959 hundreds of prematurely born babies in the United States died from the antibiotic *chloromycetin* administered as a prophylactic. In Europe babies with missing limbs and other deformities were born to pregnant mothers who took thalidomide for relief from headaches. Everywhere, even in the Arctic, mothers' milk was laced with dozens of toxic industrial chemicals.

If you don't know the horror stories behind the following names, something within you may already be dead: *Minimata disease* (mercury poisoning in Japan, 1960s), *Agent Orange* (a chemical defoliant containing dioxin used in the Vietnam War), the *Love Canal* (the upwelling of industrial chemicals into the basements of homes in Niagara Falls, New York, 1970s), *Three Mile Island* (the breakdown of a nuclear reactor in Pennsylvania, 1979), *Times Beach* (the federal buy-out of a town in Missouri heavily contaminated with PCBs and dioxin, 1982), *Bhopal* (the explosion of a Union Carbide chemical plant in India, 1984), *Chernobyl* (the explosion of a nuclear reactor in the Ukraine, 1986), and the *Exxon Valdez* (a massive oil spill off the coast of Alaska,1989). There were others.

Don't the descriptions in the last three paragraphs sound just a little bit like a modern version of Dante's *Inferno?*

What made these tragedies particularly disgusting was that most people were unaware of what was going on in their own back yards. It took Greenpeace and the media to show them. When people saw the horrors depicted in newspaper photographs and on television news, they were outraged.

Urbanites may delight in the diversified cultural and economic opportunities in cities, but shaping human society in the image of a factory is not what human evolution is all about. Coevolution with the Biosphere calls for the integration of human *Ego-Systems* with their planetary ecosystem; not between human *Ego-Systems* and their self-created technological environments. We are part of nature. Fighting nature means fighting ourselves.

For an ecosystem view of yourself and your environment, *jointly*, look out at night through the window of an illuminated room on an illuminated street. When the lights are balanced just right, you will see yourself in the middle of your personal eco-system. One half is real. The other half is reflected. The view is unique. It is yours, and yours only.

For an *Ego-System* view, turn off the inside light.

7. DISCOVERY OF THE BIOSPHERE

> Just as the hand, held before the eyes, hides
> the tallest mountain, so does our ordinary
> way of seeing hide the many wonders of
> which the world is full.
>
> Source unknown

THE BIOSPHERE IS THE living and life-supporting system in the outer part of the Earth. It is the only part of the Earth that is visible from space. It is our planetary home. It has taken a long time for members of our species to discover that we are in it.

Over billions of years the Biosphere evolved as a self-organizing system, stupendous in size, intricately complex, unbelievably beautiful, and wondrously diverse. To appreciate this, fly around the Earth in your imagination – over oceans, above mountain tops, over tropical rainforests, on Arctic tundra, and above islands and deserts. Listen to the gossip of humpback whales in the oceans. Gaze in wonder at a peacock's feathers. Watch the wind blowing through a spider's web.

For practically all of its long history the Biosphere operated without us. Nothing was known about its existence as an organized system until 1875 when Edouard Suess, an eminent Austrian geologist, gave it a name.

To understand the workings of the Biosphere the Earth had to be explored in detail, mapped and scientifically described. This involved charting the seas and continents, reconstructing the origin and evolution of the atmosphere and oceans, and mapping the distribution and abundance of different kinds of plants and animals. Minerals and rocks had to be characterized and catalogued. The global cycle of water had to be discovered. The chemical composition and circulatory patterns of the atmosphere and oceans had to be understood. Vegetational zones on land such as tropical rainforests, boreal and temperate forests, savannah, grasslands, hot and cold deserts, tundra and the like were mapped.

The idea that continents were drifting around the Earth on a sea of molten basalt was only recognized in the twentieth century. The depths of the oceans remained mostly uncharted until the twentieth century. The age of the Earth, 4.6 billion years, was not fully accepted scientifically until the middle of the twentieth century.

While all this exploration was taking place, many people believed that we had been put here by God – recently, and solely for human benefit. James Ussher (1581–1656), Archbishop of Armagh in Ireland, even calculated from Hebrew Scriptures that the Earth was created on October 23, 4004 B.C.

In contrast, another view was developing. This was passingly referred to as the anthropic principle in the preface of this book. It suggests that the entire development of the Cosmos has been predicated on the expectancy of intelligent life.

Few people had the faintest idea that we had evolved within the Biosphere, or even that there was such a thing as the Biosphere, until evolution became accepted as fact. In 1809 J.B. Lamarck provided some of the first clear evidence for evolution. In 1859 Charles Darwin, based on work that started in the 1830s, described how the process of evolution worked though descent with modification by natural selection.

In 1856, Alfred Russell Wallace, a British geographer, unaware of Darwin's work, made a strong case for geographic isolation in the origin of species. In 1867 Gregor Mendel, an Austrian monk, showed how inheritance worked. Lamarck was correct in attributing importance to habits in evolution, but he and Darwin both missed out on inheritance through genes. Mendel filled that gap.

Everything came together in the first half of the twentieth century. In 1913 Lawrence J. Henderson, a Harvard physiologist, amassed evidence suggesting that life is built into the structure of the Cosmos. In 1926 Jan Christian Smuts, of South African political and military fame, argued that the mechanism of evolution was based on the organization of matter in "wholes." In 1953 James Watson and F. H. C. Crick showed that heredity was determined by the transfer to progeny of genes made up of DNA. Serious debate over the mechanisms of evolution continues, but not on evolution as fact.

In the middle of World War I two European scientists, Vladimir I. Vernadsky, a Ukrainian-born geochemist, and Pierre Teilhard de Chardin, a French-born Jesuit anthropologist, were independently overwhelmed by discovery that humanity was changing the face of the Earth. After the war they met regularly as part a small group of interested scholars in Paris. Their purpose was to explore the scientific and philosophic implications of the concept of the Biosphere. In 1926, Vernadsky published the first book on the Biosphere. Teilhard's works were prohibited from publication by the Vatican until his death in 1956.

The atomic bomb that fell on Hiroshima in 1945 transformed the Biosphere from a little-known intellectual concept into a geochemical and political reality. Politicians knew that an arms race would follow, but neither they nor any of the scientists, engineers and mathematicians who constructed the bomb had the slightest idea of the global ecological effects that would ensue. The notion of ecology had not penetrated their minds.

In the ensuing arms race radioactive isotopes rained down from the stratosphere all over the Earth. In the Arctic, radioactive iodine was concentrated in the thyroid glands of reindeer. In the middle of the Pacific Ocean, radioactive cobalt turned up in giant clams. Radioactive strontium penetrated the teeth and bones of children everywhere on Earth.

In 1972 James Lovelock, British inventor and scientist, suggested that the Biosphere could be a self-regulating system, even changing the climate of the Earth to enhance its survival. He called this "the Gaia Hypothesis" after Gaia, the ancient Greek goddess of the Earth.

Much remains to be learned about the workings of the Biosphere, and the same is true for ourselves. We have yet to discover all the details on how our species came into being, our relation to close relatives such as Neanderthal man, how our bodies work, the origin of languages, the migrations of our ancestors, and the workings of the human brain. In reflecting on what is known and not known, keep in mind that ecology is like a photograph; evolution, like a moving picture.

Humanity has been slow in coming to recognize the importance of the Biosphere as a supranational political concept. An almost unbelievable planetary sequence of crises may be indispensable for our enlightenment. Then, possibly, we may, perhaps begrudgingly, make a positive decision to co-evolve with the Biosphere instead of trying to fight it.

8. THE BIOSPHERE AS AN
ORGANIZED SYSTEM

There is no reality to find beyond the one
that you make by looking.

Werner Heisenberg

⁓ IN THE WORDS OF the Greek poet Hesiod
(eighth century B.C.) we find the fair Earth giving birth to the starry heavens as an entity equal
to herself. In the Greek myth of the Garden of the
Hesperides, the nymphs nurture the golden apples
bestowed as a wedding gift by Gaia, Mother Earth,
on Hera, goddess of women and marriage.

What wonderful images these words create!
But except for Lovelock's use of the *Gaia hypothesis*
to refer to the Biosphere as a self-regulating system,
myths have been mostly bypassed or abandoned in
the secular and pluralistic scholarship of the mod-

ern world. *Biosphere* is now the accepted scientific name for Mother Earth.

The Biosphere consists of:

- the *atmosphere* (78% nitrogen, 21% oxygen, 1% argon (an unreactive gas), nearly 0.04% carbon dioxide and rising, and variable amounts of water vapor)

- the *hydrosphere* (rivers, lakes, oceans, and ground water)

- the upper part of the *lithosphere* (soils, sediments, rocks, and permanently frozen ground) and

- between three million and thirty million species or more of living organisms.

That's a mighty big system! Everything in it is functionally connected. It has been operating for billions of years. If you think that we can fight it and come out on top, you might want to think again.

Reconstruction of the history of the Earth has taken centuries of cooperative scientific effort. The current version, subject to revision, is riddled with all sorts of hidden assumptions, speculations, and unanswered questions. It is like a jigsaw puzzle with oceanic expanses between little islands of assembled pieces, and without detailed photographs to show what the puzzle looked like at any particular time. Nevertheless, enough pieces have been fitted together to make a story that is stranger than fiction.

Cosmologists say that billions of years ago, the Universe exploded into being from a speck of "nothing". Within minutes atoms of hydrogen and helium formed. Galaxies burst out of nowhere, flying in all directions. Billions of years later a first generation star in one of the galaxies ran out of fuel. First it imploded, then it exploded, spewing out an enormous cloud of dust and gas. As the dust cloud expanded it cooled, and as it cooled it contracted into a giant whirling disc. Out of the disc came a second generation star and its planets. That star was our sun. Our Earth was one of its planets.

Astronomers say that the Earth accumulated from rocks of various shapes and sizes over millions of years. Originally, so it is said, the Earth was hot, lifeless and without an oxidizing atmosphere. Over time, a secondary atmosphere arose from enormous burps of steam and hot acidic gases spewed up from the Earth's molten interior. Comets crashed into the Earth, enriching the planet with water.

Geologists say that rains gave rise to rivers, lakes and oceans. The great water cycle of the Earth came into being. Oxygen accumulated in the atmosphere as a waste product of photosynthesis. The presence of oxygen in the atmosphere increased tenfold the energy that organisms could obtain from food. Large organisms evolved. Life expanded from water to land. Trees, dinosaurs and primates appeared on Earth.

Compressing the 4.6-billion-year history of the Earth into a week, paleontologists say that unicellular life arose before noon of the second day; photosynthesis by noon of the third day; and a mildly oxidizing atmosphere by the fifth day. Late on the sixth day, large, multicellular organisms burst forth on the scene. On the seventh day, trilobites and fish appeared in the morning. Dinosaurs came and departed in the afternoon. Birds and mammals appeared in the evening.

Our species strode onto the stage in the final minutes of the week. Languages evolved during the last few seconds; agriculture, in the final second; and you popped out of your mother's womb in the last hundredth of a second of the Earth's first week.

9. TEN PRECIOUS LEGACIES
LEFT BY THE BIOSPHERE

He who loves the world as his body may be
trusted with the empire.

Lao Tzu

⤳ THE ARGUMENT IN THIS chapter begins with
you. If you believe that all systems are connected
– the glint of sunlight caressing a drop of dew on a
blade of grass, the incessant breaking of waves on
the shores of a great ocean, the wind playing tag
with itself between the leaves of an old oak tree,
the contrast of grey limestone, pink granite, and
red sunsets – if you credit co-dependence and interconnectedness between your body and the Earth,
then it follows that you are a child of the Universe.
You have a right to be here. With that right comes
an obligation to honor and protect the planetary
heritage left by your kin.

By planetary heritage, I mean the living-and-life-supporting system of the Earth that we call the Biosphere. By kin, I mean the innumerable forms of life that over incredible expanses of time have lived on Earth. The message of the tree of life, now amply documented by DNA, is that all organisms are kin – distant kin for the most part, but nevertheless kin.

Our biospheric kin have left ten precious gifts, legacies, endowments – whatever you want to call them. These are:

(1) oxygen in the air we breathe – which also created the ozone layer that protects us from harmful ultraviolet light

(2) the food we eat, and the food webs of other organisms which, by transforming wastes into resources, keep the Biosphere fit for life

(3) the water we drink and bathe in which nourishes all forms of life

(4) reproductive cells that miraculously transform fertilized eggs into adult beings and then back to fertilized eggs

(5) the "lowly" microbes in our guts without which we could not live

(6) the bodies of former plants and animals transformed into deposits of natural gas, petroleum, and coal that permitted the rise of technological civilizations

(7) organisms and their ecosystems that, like miners' canaries, provide warning signals of the dangers to which we are exposing ourselves

(8) wood to make chopsticks, build homes and cook food

(9) the beauty of nature that brings peace to our minds, and

(10) the continuity of the Biosphere which, over billions of years, enabled our species to flower on the tree of life.

In turn, the Biosphere depends on the Earth as its physical support base. It relies on the sun for the radiant energy that it needs to run and on the various forces of nature, including gravity, that maintain the unity of the Cosmos as a whole.

Most scientists claim that there was no intentionality in all this. They say that it was simply "the way that things worked out." If so, we do not need to feel bound by any indebtedness to the Universe or our ancestors for our existence.

And yet one wonders. Can it be accidental that we are here? Could there be some hidden cosmic meaning in the legacies left by the Biosphere?

Natural selection is the sieve that is responsible for adaptive change. It is anything but accidental. What is accidental is the specific direction that evolution has taken in particular cases. Most of that direction appears to have come from organisms "muddling through". Over and above that, there may be a grand design at work in the Universe, but what that design might be and how it works remain a mystery. But one thing is clear: evolution does not just pertain to living organisms. Organisms and their environments evolve jointly.

Teilhard asked: "Could there be man without Earth?" He knew that the answer had to be *no*. The forces of natural selection dictate against this. Our bodies are adapted to living on Earth. Gravity reminds us, day and night, that we live on Earth. Take a flight in space if you don't believe it. Or experience jet lag. DNA confirms our unity with the rest of life. To live in any other part of the Universe we would have to take a "Noah's arc" full of the Biosphere, and more, with us. We should therefore be very thankful for the legacies left by the Universe and Biosphere.

PART III

HUMAN UTOPIA: DREAM OR NIGHTMARE?

> Under a good administration, the Nile gains
> on the desert. Under a bad administration,
> the desert gains on the Nile.
>
> Napoleon Bonaparte

⌇ THE TWENTIETH CENTURY VIEWED as an era of global ecological madness. Upward causation and downward causation described and related to a change of context. The *Sorcerer* as a life-force at work within us. The runaway cycle of human technology and population. Demotechnics explained. The three evils of technological civilizations. Experiencing global stress by blowing up an Earth balloon. How systems change. The difficulties of changing the demotechnic system. How humanity is preparing the ground for a planetary glut of crises by trying solve human problems without curtailing runaway growth. Five long-term crises that could bring an end to human technological civilizations in the next 500 to 1,000 years. How the overgrowth of human technology and population may be leading human civilization toward coevolution with the Biosphere.

10. ECOLOGICAL MADNESS

> "If the development of civilization has such
> a far-reaching similarity to the development
> of the individual and it employs the same
> methods, may we not be justified in reach-
> ing the diagnosis that, under the influence
> of cultural urges, some civilizations, or some
> epochs of civilization – possibly the whole of
> mankind, have become "neurotic?"
>
> From *Civilization and its Discontents*
> (1930) by Sigmund Freud.

FREUD HAD THREE CONCERNS with attempts to answer the unsettling question that he posed:

(1)   that the relationship between personal and communal neuroses might be based more on analogy than on common causes

(2)   that communal diagnoses lack the experimen-

tal controls necessary for rigorous analysis, and

(3)   that no one has the authority to practise communal therapy, if and when needed.

One cannot argue with Freud's first concern. Analogies can be confusing. Reasoning based on common causes is a more reliable foundation for making diagnoses, connections and predictions.

In regard to Freud's second concern, experimental controls really apply only to simplified laboratory situations. In complex systems the ability to change one thing at a time is rarely, if ever, possible. But while the lack of experimental controls does complicate analysis, it does not make rigorous analysis impossible.

In regard to Freud's third concern, that no one has the authority to practise communal therapy, the fact is that the whole world provides it when needed. Wars, even though painful and often unproductive, can lead to changes affecting the behavior of individuals and nations. Some significant innovations that followed World War II were: enactment of laws against crimes of hate, creation of the United Nations Organization, and plans for a European Union. Therapy doesn't have to be practiced on a psychiatrist's couch.

Respected analysts of human societies in the twentieth century persisted in asking the same question that Freud posed. These included Lewis Mumford (*In the Name of Sanity* 1954); William Leiss (*The Domination of Nature* 1972); Paul Shepard (*Nature and Madness,* 1983); and numerous articles and political commentaries in *The Nation, The New Republic,* and other American magazines during the Cold War.

The popularity of the magazine *MAD* and films such as *The Day the Earth Stood Still* (1951), *Dr. Strangelove (1964), Seven Days in May* (1965), *Planet of the Apes* (1968), *The Gods must be Crazy* (1975), *The Island of Dr. Moreau* (1977, 1996) and *The Postman* (1997) show that the madness was evident on several levels of human understanding.

The twentieth century was a time of great discoveries and inventions. These ranged from DNA as the language of genes to aeroplane flight, radioastronomy, antibiotics, television, nuclear energy and transistors. But in ecological retrospect the twentieth century can also be seen as an era of global madness: two world wars with an intervening global depression, six million Jews paraded naked into gas chambers, the escalating scale of pollution, the rain of human blood percolating through the soils of Africa, the failure to eliminate poverty everywhere, the Arms Race, the Cold War, the Vietnam War, the Berlin Wall, the spread of HIV/AIDS, and others.

To attribute all the dysfunctional events of the twentieth century to a single cause would be absurd; but if one cause had to be identified above all others, the runaway growth of human technology and population, jointly, would be a good candidate. Rising levels of technology and population on Earth act like rising temperatures and pressures in a chemist's flask. Emotions such as hate, greed, and the lust for power, raise the rates and intensities of human interactions by lowering the activation energy needed to bring them about.

One does not have to be a psychologist to know that the Faustian dream of dominating nature is ecological madness. There is also the economic myth of a person who sees all, knows all, and acts rationally – when, in fact, no such person exists. The illusion that the intertwined growth of human technology and population can go on indefinitely shows that humanity is on a suicidal course. It is a dangerous fantasy to see yourself as separate from nature. It is ecological madness, pure and simple.

The subversive influences, conscious and unconscious, at work behind this madness are the cravings that lead to The Seven Deadly Sins.

St. Gregory the Great categorized The Seven Deadly Sins in the sixth century A.D. as: lust, gluttony, avarice (covetousness, greed), sloth (idleness, waste of time), wrath (anger, hate), envy (jealousy). and pride (vanity). In 1589 Peter Binsfeld named the demons behind the Sins as: Asmodai (lust), Beelzebub (gluttony), Mammon (avarice),

Belphegor (sloth), Satan (wrath), Leviathan (envy), and Lucifer (pride). Now, by the stroke of a pen, I identify the mastermind that controls the demons as: the Sorcerer. Identifying the Sorcerer as the mastermind behind the cravings could be key to controlling the sins.

Giving in for a moment or a lifetime to the urges, cravings and other excesses brought on by the Sorcerer is far more crippling than doubts. The opposites to the Seven Deadly Sins are the Seven Cardinal Virtues: humility, meekness, charity, chastity, moderation, zeal and generosity. The question is: Can the Virtues overcome the Sins?

We have been programmed by evolution to ensure survival by excessive reproduction in early phases of population growth. In later phases of population growth, ecological limits come into play. Excessive reproduction, once useful for survival of the species, becomes injurious to the continuity of life when continued belong its useful time.

The context of life on Earth has changed, and the Sorcerer apparently does not know it. More on the Sorcerer will unfold in the last section of this book. For now, it is enough to say that we have been, and continue to be, the Sorcerer's apprentices.

Are greed, envy, deception and desire what progress is all about? Is growth really an end in itself, independent of the context in which it takes place? How much further is there to go? Is it too late to change direction? Are we on an irreversible path?

To understand answers to these questions it is essential to grasp how much the context of human life on Earth has been transformed by the runaway growth of technology and population.

11. A CHANGE OF CONTEXT

> We tend to think that revolutions are the
> cause of change. Actually, it is the other way
> around: change prepares the ground for
> revolution.
>
> from *The Temper of our Time*
> by Eric Hoffer

⤳ HUMAN CIVILIZATIONS HAVE DISRUPTED the wholeness of the Biosphere in innumerable ways. Our transgressions against nature have included hunting large mammals to the point of extinction, overexploitation of fisheries, drawing on groundwater until wells run dry, turning lush and productive grasslands and forests into deserts, polluting air and water, thinning the ozone layer, and changing the Earth's climate by creeping additions of greenhouse gases to the atmosphere.

Now, in an age of declining material growth,

we are being forced to change our life style from parasitism on the Biosphere to coevolution with the Biosphere. Parasitism on the Biosphere means drawing on the Biosphere's capital without giving anything in return. That is like killing the goose that laid the golden eggs. Coevolution with the Biosphere means evolving in harmony with nature similar to the ways in which many flowers and insects have coevolved to serve the other's interests in pollination or making honey.

The shift from parasitism on the Biosphere to coevolution with the Biosphere is being forced on us by the interaction of two complementary opposites that, together, have driven the process of evolution over time. These opposites are: the tendency of populations to overproduce in relation to the resources needed to support them, and the role of the environment in cutting populations down to size. The change is from an era of *upward causation* to an era of *downward causation* as shown in Figure 11.1.

To grasp the meaning of these terms consider the growth of micro-organisms in a container provided with all the nutrients they need to grow. Their numbers rise exponentially, doubling over constant or even shortening intervals of time. The cause of the increase lies in the power of living organisms to reproduce in favorable environments. That is an example of *upward causation.*

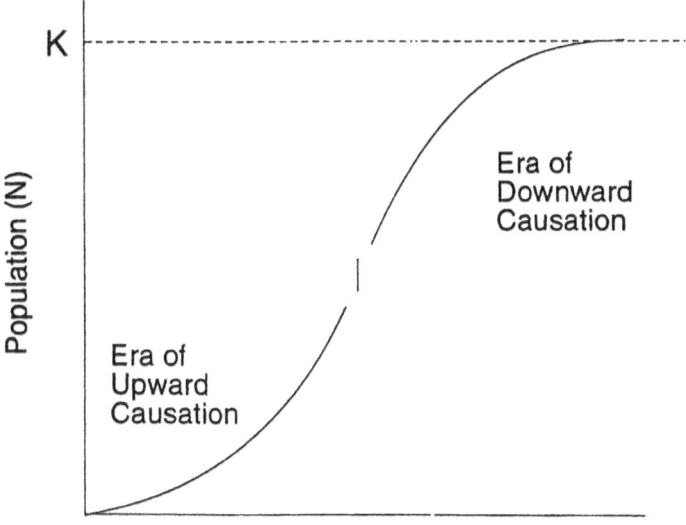

Time

Figure 11.1. The S-shaped curve of ecological growh divided into eras of upward causation and downward causation. See text for explanation. N refers to the number of organisms in the population. K refers to the carrying capacity of the ecosystem for a population. Note how constraints to population growth appear long before the carrying capacity under a given set of conditions becomes apparent.

Over time, the micro-organisms become so numerous that they change the environment inside the container. Nutrients decline, wastes accumulate, and life becomes increasingly difficult. Deaths increase and growth grinds to a halt. That is an example of *downward causation*. That is how things generally work in the natural world.

In the human world, Faustian beliefs and practices take over. The human penalties imposed by downward causation are postponed insofar as possible. Population increases seemingly without limit. When the population moves across the balance point between upward and downward causation it can be said to have entered a *forbidden zone*. Death rates rise sharply the more deeply the population penetrates into the forbidden zone.

The significance of the container is that it forces everything within its boundaries to interact. Catalysts change the field of selection that governs the balance of life and death. In this way downward causation influences the course of future events.

Before measures were introduced to limit the spread of infectious diseases such as malaria, typhoid fever, cholera and dysentery, half of the children born died before reaching the age of five. The average length of life from birth was under forty years. After measures were introduced to control the spread of disease, population increased because of the prolongation of human life.

The general rule is: when the growth of one part of an integrated system threatens the integrity of the larger system that contains it through upward causation, the larger system fights back with downward causation. In this sense, it is metaphorically accurate to speak of the revenge of the Biosphere in return for human excesses.

It is the same in principle for the membranes that enclose cells, the skin that encloses the body, the interactions of chemicals inside a reaction vessel, and the upper and lower boundaries of the Biosphere. In this sense, a fair share of all the tumultuous social, economic and political upheavals of the twentieth century can be attributed to downward causation from the Biosphere..

The reason for describing the exponential growth of human technology and population as ecological madness can now be seen: *It is the result of continuing behaviors and practices that were adaptive in an era of upward causation, but are suicidal when continued in an era of downward causation.*

Call it the growth syndrome and you will understand it at once. Side effects include stresses on children that can persist throughout life, the decay of urban cores, injuries from traffic collisions, pollution and other tragic events associated with the rapidly escalating pace of modern life. Little wonder that the acceleration of growth has fragmented people's minds and, as a result, has disrupted the wholeness of nature.

Surely, we must reject this suicidal behavior in Human Utopia as madness. Of course, the "extinction" of Human Utopia does not necessarily imply extinction of the human species. Nevertheless, the change of context brought about by the runaway growth of technology and population has prepared the ground for revolution. The only intelligent questions now are: How much time is there for us to slow down before a dark age ensues after the inevitable crash? What strategies can we adopt to redirect the pace of runaway growth of human technology and population in ways that nourish the Biosphere and ourselves? What are the long-term effects likely to be of the teenage obsession for instant gratification?

But most people are not asking these questions. Postponing the long-term consequences of an already fated system is what really makes our current way of living madness. The Hopi Indians of the American southwest have a word for it: *koyaanisqatsi* – meaning, out of balance, a state of life that calls for another way of living.

12. THE RUNAWAY DEMOTECHNIC CYCLE

> Until the crisis validates itself by catastro-
> phe, the whole concern is an abstraction, in
> the sense of not entering actively into our
> consciousness, its dreams, fears, fantasies.
>
> Richard Falk, *This Endangered Planet*

⌣ To EVALUATE THE BIOSPHERIC impact of different species on the basis of head counts alone
is like comparing dinosaurs to mice based on the
numbers of individuals. Our species is an intimate
blend of flesh and technology, and that technology
is very powerful. For that reason the primary focus
of this chapter is on metabolism, not head counts of
populations. The accent is on technologies that rapidly spread from person to person and culture to
culture. External energy supply is a good example.
It started with wood, wind and water power; and

then led to coal, electricity, natural gas, petroleum, geothermal energy and nuclear power.

Contrary to popular belief, technology (often better referred to as "technique" or "know-how") does not just refer to things. It also refers to processes, including biochemical cycles and the "know-how" involved in thought. Nor is technology limited to our species. Beavers build dams and houses. Bees build hives and birds build nests. Deep-sea lantern fish dangle lights in front of their mouths to lure prey. Woodpecker finches use cactus spines as cocktail picks to spear insect larvae under the bark of trees. Lions hunt their prey in teams.

Living organisms use food to provide the energy needed to drive biochemical transformations in their bodies. This process is called metabolism. It is like fire without a flame. Organic chemicals are broken down to yield little packets of energy-rich compounds that are distributed to cells in various parts of the body where they are used to synthesize proteins, contract muscles, and produce heat.

The ability to control external energy is one of the most remarkable features of our species. It has allowed us to benefit from both external (technological) metabolism and internal (physiological) metabolism. So the next time you look at a fellow human being walking down the street do not measure the power of that person by what you see. Look at the person as an enormous bundle of invisible energy, most of which is external and technological. Our problem is that we don't see the connections.

The runaway growth of human technology and population, jointly, has changed the context of life in the Biosphere from an era of upward causation to an era of downward causation. Much of that growth has passed unnoticed, partly because of the slowness of its change from year to year, and partly because of the lack of a word linking the invisible connections between technology and population.

Words are forms of technology. An example of what can happen when a word is missing from our vocabulary came in the 1970s when two prominent American ecologists, Paul Ehrlich and Barry Commoner, became caught up in a public debate on the causes and cures of the environmental crisis.

Ehrlich maintained that overpopulation was the primary concern because of its insidious long-term effects and the lack of controls. Commoner argued that uncontrolled growth of socially harmful technologies deserved more attention because of their amplifying power and relative ease of control.

Both knew that the growth of technology and population, *jointly*, had caused the environmental crisis of the 1960s. However, lacking a word to express the interaction, they left a false impression that the public had to be exclusively concerned about one or the other.

Not so.

Human population and technology have been chasing each other globally in a never-catch-up, runaway cycle for centuries. By easing the conditions of human life, technology spurred the growth of human population. Then the increased population demanded and usually got more and faster, if not "better", technology.

At first, the rates of runaway growth doubled as a single exponential over constant intervals of time. Then, in the nineteenth century, the rates began to rise as a double exponential over shortening intervals of time.

A runaway cycle like that doesn't make sense in the context of bounded systems such as Mouse Utopia and the Biosphere.

Ecological balance demands a two-way interaction in which more of one leads to less of the other, and less of the other leads to more of the first. That is how complementary parts work together in our bodies – in muscles, hormones, and temperature regulation. It is the same for the Biosphere.

To give reality to the interaction of technology and population, I coined a new word in 1972. The word was *demotechnic* (from the Greek: *demos*, population; *techne*, technology). *Demotechnics* is the noun; *demotechnic*, the adjective.

In 1978, I showed how to express the combined technological and physiological energy consumption of human populations quantitatively in demotechnic units. Just as a "standard" horse is needed to define one unit of horsepower, so is a "standard" human needed to define one demotechnic unit (or D-unit for short) of energy consumption.

One D-unit is defined as the food energy required to maintain an average human for a year at a rate of 2,333 kilocalories of food energy per day. This was the daily requirement per person in 1988-90 for medium development countries cited in The Human Development Report for 1993. Based on that definition, physiological energy consumption (P) in D-units is numerically equal to population counts in heads. Technological energy consumption (T) (usually measured in joules) can be converted to D-units on the basis that 1 kilocalorie is equivalent to 4,190 joules.

Finally, the demotechnic index (T/P), or D-index for short, is a rough-and-ready index of affluence. Currently the global average D-index is about 16, ranging from less than 1.0 for some countries in Africa to about 120 for Canada.

Using appropriate conversion factors it is a simple matter to combine technological (T) and physiological (P) energy consumption, individually, nationally and globally, as T + P. In 2000, the five most voracious energy gluttons of the Earth were, in decreasing order: the U.S.A., the former U.S.S.R.,

China, Japan, and Germany. The United States ac-
counted for more than one quarter of the Earth's
total human energy consumption in 2000. Early in
the new millennium India bumped Germany off
the list.

Table 12.1 compares the global rates of increase of human population in heads to total energy
metabolism in D-units between 1950 and 2000. Note
how rapidly total energy consumption increased
relative to population counts in heads. Also, note
the extent to which the total energy consumption in
1950 grossly exceeded what would have been predicted from population counts in heads, and how
the global average D-index has been leveling off
since 1970.

Don't run away!

Two examples show the ecological absurdity
of viewing population and technology separately:

(1) *Pollution in Lake Erie.* Between 1800 and 2000,
     human population in the drainage basin of
     Lake Erie increased one hundred-fold; and
     technological energy consumption per person
     increased another hundred-fold. The overall
     demotechnic increase was thus ten thousand-
     fold in two hundred years. Little wonder that,
     on October 20, 1961, when American fisher-
     ies biologist, Stanford H. Smith, commented
     publicly that "Lake Erie is dying" it became
     front page news.

(2)   *Global Demotechnics.* The human population of the Earth rose from 3.9 to over 6.0 billion people between 1973, the year of the oil embargo, and 2000. The increase was stunning when first predicted in 1973, but nowadays all it gets is a shrug of indifference. *More significantly, the global increase between 1973 and 2000 expressed in D-units rose from an equivalent of 67 billion nontechnological people in 1973 to 106 billion nontechnological people in 2000.* That is an awful lot of nontechnological people!

Year	P	T	T + P	T/P
1950	2,519	19,750	22,269	7.8
1960	3,020	32,590	35,610	10.8
1970	3,691	52,786	56,477	14.8
1980	4,430	70,172	74,602	15.8
1990	5,255	84,437	89, 692	16.1
2000	6,075	99,643	105,700	16.5

Table 12.1. Annual global energy statistics. Human mid-year physiological energy consumption (P), technological energy consumption (T) and total human energy metabolism (T + P). All numbers are expressed in millions of D-units except the D-index (T/P) which is a dimensionless ratio roughly equivalent to affluence. See the text for further explanation.

Having made the point that our species is demotechnic in nature, the word *demotechnic* can be skipped in most instances. Thus, *total energy consumption* for a human population will mean *demotechnic energy consumption*, the *runaway demotechnic cycle* can be shortened to *runaway cycle*, and *demotechnic growth* can be shortened to *growth*.

The rates of growth of total human energy consumption are now so high that they are endangering the health and integrity of the Biosphere and the viability of our species. We are in the forbidden zone. Time is shortening. Space is closing in.

In 1962 Rachel Carson warned in *Silent Spring* that long-lived chlorinated hydrocarbons such as DDT and DDE were not just killers of insects. They were killers of life. Her landmark best-seller was followed in 1972 by The Club of Rome's best-seller *The Limits to Growth*. In 1992 came *The World Scientists' Warning to Humanity*, signed by more than 1600 leading scientists, including 104 Nobel Laureates and learned institutions in 70 countries.

None of these chilling indictments had the slightest effect in restraining the craving for growth of most economists, businessmen, industrialists, entrepreneurs and politicians. Growth is still subsidized on the false belief that further growth will bring wealth and happiness to most people. In fact, the reverse is more likely to be true. Further growth will only increase the power and wealth of those who make money from the bonanza. The runaway growth cycle is not an illusion. The illusion is our piecemeal belief that human technology and population are best viewed independently. The disastrous global effects of the runaway growth cycle have not actively entered our consciousness.

Lester Brown (2005) has suggested why the western economic model may lead to the downfall of human civilization. In one word, it is mimicry - by China and India of the American economic dream. On the basis of current projections it has been estimated that China will likely overtake the United States as economic leader of the world by 2030, and that India will not be far behind.

Piecemeal thinking in an era of downward causation can be disastrous. One example is the proposal to increase nuclear energy globally in the hope of reducing global climate warming. In fact, reactors produce carbon dioxide in their construction and maintenance; they give rise to radioactive wastes that cannot be recycled; their construction diverts funds away from more ecologically sound projects; and they are prime targets for terrorists. In an era of downward causation, energy conservation is the only answer.

13. THE THREE EVILS OF
TECHNOLOGICAL CIVILIZATIONS

"Evil be thou my good."

Satan in Milton's *Paradise Lost*

 "ONLY THREE EVILS? ABSURD!"
Of course, you are right. There are hundreds
of evils. The exact number depends on one's point
of view. My aim here is to accent three primary evils
that set the runaway cycle in motion and perpetuate its direction and runaway rate.

The three primary evils of technological civilization based on demotechnic reasoning are: money,
the clock, and forms of advertising and propaganda
based on deception. Controlled fire might be added
as a fourth evil, however it is an evil of a different
sort. It is an absolute essential of any technological
civilization. The other three evils can to some extent
be bypassed or done without.

Each of the three evils is a human invention. In that sense one might say that there is only one evil: the process of invention itself. But that would imply that evolution is inherently evil, for what is evolution but invention glorified? It would be like asserting that the entire Universe is inherently evil – galaxies, stars, planets, plants, animals, the whole lot. There wouldn't be much sense in that.

Coinage was invented in Greece some time before the sixth century, B.C. Since then money has debased all human values by making those values interchangeable. Money can be used to buy political favors, to arrange murders at will, and to treat land as a commodity rather than as a sacred trust. When money becomes the dominant category of thought all human values, including moral values, come to be reckoned in money. By providing credit for ecologically disastrous mega-projects, bankers and other dispensers of venture capital jeopardize coevolution with the Biosphere.

Water clocks were invented in antiquity, but it was only in the fourteenth century when *mechanical clocks* made their way into bell towers in European cities that commerce began to march to a common beat. As mechanical clocks spread from city to city, and nation to nation, the demotechnic cycle became global in scope. When that happened, the regimentation of humanity was complete.

Deception with a view to gaining an end is not a purely human trait. It exists in a variety of animals, including primates. Camouflage, leading predators away from a nest, and playing dead are three common examples. Most of these natural forms of deception are used to enhance survival of the population or species. In contrast, deception in human societies is mostly used to enhance individual or corporate power, or to make money, or both.

Following Jacques Ellul (1911-1994), French author of *Propaganda* (1954), it is useful to distinguish two forms of propaganda – sociological propaganda that reinforces a common life-style such as "the American way of life", and propaganda that leads its audience in a preferred economic or political direction. Sometimes, as in the specious logic of television, it is difficult to know where mass communication, advertising, and propaganda begin and end. The key element in respect to the runaway growth cycle is *deception*.

Words and snappy phrases are commonly used as vehicles for deception in our species. Combined with our propensity for piecemeal thinking, they lead us into believing that for every ailment there is a pill to cure it. Television ads for medicines often end with "ask your doctor" or "recommended by doctors" as if to suggest that doctors fully endorse the advertiser's case.

How can the three evils be combated? By

changing the uses to which they are put. All that is needed is to turn the "three evils" into "three goods". Instead of using money to degrade the Biosphere, it could be used, as some of it is used today, to nurture the health and integrity of the Biosphere. Instead of using land for human purposes land could be set aside, as some of it is today, for conservation and the protection of endangered species and ecosystems. Instead of corrupting educational systems by encouraging them to ritualize and perpetuate the craving for growth, educational systems could help young people to adopt less predatory approaches to living in the Biosphere.

These are but three of hundreds of ways in which the health and integrity of the Biosphere could be enhanced. If people subsidize the direction in which they want their society to move, their society will move in that direction.

Extend the golden rule (*Do unto others as you would have others do unto you*) to the golden ecosystem rule *(Do unto the ecosystems that you share with others as you would have others do unto the ecosystems that they share with you)* and there could be the makings of a moral planetary civilization.

It is unlikely that the three primary evils can be eliminated from human societies, but they can be made to follow more ecological guide-lines and rules. Out of these guidelines positive propaganda for the Biosphere could emerge. By intelligent transformative processes the three evils could become our common good.

However, the old questions return:

Can we? Will we? In time?

14. EXPERIENCING GLOBAL STRESS

> You never know what is enough unless you
> know what is more than enough.
>
> William Blake

⁀ IF YOU HAVE FOLLOWED my argument this far, you will realize that we have reached a hard stumbling block to implementing plans and policies for coevolution with the Biosphere. How can people know what is enough in the Biosphere, when they do not know what the Biosphere is, let alone how it is increasingly stressing them from downward causation brought on by the runaway growth of technology and population?

H.G. WELLS PUBLISHED his version of Human Utopia and how to attain it in *The Open Conspiracy* (1928). He called for a planetary unification of human purpose. Unlike most utopias his was based on a clear understanding of limitations imposed by human behavior and evolution. What he lacked was knowledge of ecology. Wells died in 1945, only a few years before the environmental crisis of the 1960s when the public sense of ecology came into being.

When Wells reached a point in *The Open Conspiracy* that was beyond the power of reason to explain he invented a parable about three sailors and a cabin-boy who were shipwrecked on a deserted island. One day they discovered a pig on the island. Immediately, each devised a plan to catch his favorite part of the pig. The cabin-boy (H. G. Wells) proposed building a trap to catch the whole pig, but the others ridiculed this as far too ambitious a plan. Thus, each went his own way and in the end the pig ran free and the sailors starved until picked up by a passing ship.

The moral of Wells's parable was that a partial enterprise is not always wiser or more hopeful than a comprehensive one. His advice to participants in the open conspiracy was: "Go whole hog!"

Wells's parable was a "soldierly" attempt to persuade his readers to opt for human unification, but it didn't stir people emotionally. It was too far off and too logical. It called for cooperation and it didn't reveal the magnitude of the undertaking.

In 1980, I created a mythical character named *Johnny Biosphere* to remind young people that we live on Earth. Johnny's first words of greeting are:

"Hi! I'm *Johnny Biosphere*. I come from Earth. Where do you come from?"

Johnny converses with young people about the Earth with a globe on his back. His stories are fun and they pack a deep ecological message.

To show the catastrophic effects of runaway growth Mr Biosphere blows up an Earth balloon until it bursts. The thickness of the balloon in relation to his globe is directly proportional to the thickness of the Biosphere in relation to the Earth. The balloon is easy to relate to. Fully inflated, it is about the same size as an adult human head.

Mr. Biosphere asks students in schools to raise their hands if they want more people, more cars, more television sets, more candy bars, and more bubble gum on Earth. Then he says, "Hands down!" and asks those who vote for no more material growth on Earth to raise their hands.

If the growth people are in the majority *Mr. Biosphere* blows more air in the balloon and takes another vote. If the no-growth people are in the majority, the biosphere balloon goes free.

Before 1988, nine times out of ten, the growth advocates won out. With successive votes the balloon got bigger and bigger. As the balloon neared the bursting point students cringed and distanced themselves from the balloon. When the balloon popped, they all roared with laughter and fell over dead. They wanted the balloon to pop.

After 1988, it was the other way around. People, young and old, throughout the world voted to save the Biosphere. Evidently, a vital connection had been made between their minds and the Biosphere. It has been the same everywhere ever since. Something changed in human minds in 1988. David Suzuki, talented Canadian scientist and showman, was one of the first to note it.

Photographs of the Earth from space in the 1960s set the stage for this new consciousness. The report of a hole in the ozone layer over Antarctica in 1985 dramatized the seriousness of the threat. Most young people learned about the hole while channel-flicking on TV. They interpreted the hole as a signal of "game over" on Earth.

The bottom-line came in 1988, the year of the hot, dry summer in North America. That summer you couldn't buy an air conditioner for love or money in most of North America. People standing at street corners and sitting on park benches were talking about global warming and the hole in the ozone layer.

I like the balloon story because of the way that people cringe when the balloon is about to burst. People intuitively know that any weak point on the balloon's surface can cause the balloon to go pop; and they also know that there are lots of "weak points" on Earth.

But do people feel the strain of runaway

growth when trying to cross a busy downtown street, when yearning to buy that new car in the showroom, or when choosing to give birth to another child?

Maybe, but for many people the connection with the Biosphere is still remote. Most people feel helpless. They know that they, alone, cannot change the destructive aspects of the human system. And from that they conclude, wrongly, that there is nothing they can do.

William Blake was right. You can never know what is enough until you know what is more than enough; and when you experience what is more than enough, nobody has to tell you what to do.

If we don't start paying more attention early on to the distress signals of the Biosphere, such as parentally induced stresses on children, endangered species, endangered ecosystems, the hole in the ozone layer, and global climate change, it could be too late to reverse the consequences of humanity's previous indifferent actions.

15. RESISTANCE TO CHANGE

> In nature there are neither rewards nor pun-
> ishments - there are only consequences.
>
> Robert Green Ingersoll, *Some Reasons Why*

WE HAVE NOW REACHED another crucial stumbling block to implementing plans and policies for coevolution with the Biosphere. In the last chapter the stumbling block was the difficulty of directly and knowingly experiencing global stresses from the runaway growth of technology and population. In this chapter it is the difficulty of modifying the countless ecologically dysfunctional connections that humanity has set in place to spur the runaway global demotechnic cycle on its destructive course.

The saying "You can't change City Hall" shows that most people know how resistant systems are to change. However, resistance to change does not mean incapable of change. It can be the prelude to a shift of interactions to a higher level of integration. Detailed scientific understanding of the behavior of systems only began in World War II. It came from the development of information theory and the control of guided missiles.

H. G. Wells (1928) was right in suggesting that partial enterprises are not always wiser or more hopeful than comprehensive ones, but he underestimated the global complexity of the demotechnic system that he needed to change. Also, the types of skilled participants that he wanted to attract to the open conspiracy were often the very same as those who had caused the degradation of the Biosphere to take place.

Both the Biosphere and the global demotechnic cycle are large, overlapping, interlocking, and highly organized systems. They resist change; and when they do change it is often by "flipping" or "flickering" into states that reconnect in strange new ways with neighboring systems.

The alternation of glacial and interglacial epochs over the past few million years is a global illustration of how systems commonly change. There may be hints of what is coming, but not the final shock of surprise when a continental glacier bulldozes your home off the map. Flicker, flicker, flip, flop! It is similar for many aspects of evolutionary change.

The rash of air and water pollution that spread

over the face of the Earth in the 1960s was a global flip. Pustules of pollution erupted in areas of dense technology and population. Never before had the Earth descended to such an unhealthy state. The rash was reduced through pollution abatement programs, but pollution was not eliminated. The craving for growth was still too deeply embedded in the structure of human societies.

In Lake Erie, a flip occurred during three days of abnormally still weather in September, 1953. The still weather choked off the vertical circulation of water, with the result that dissolved oxygen disappeared from the bottom waters of the western and central basins of the lake. Enormous populations of mayfly nymphs, important items in the diets of fish, died overnight. The human cause was a long history of water and sediment pollution. Only now, half a century later, are the mayflies beginning to return.

On a global scale, here are some of the interlocking technological subsystems that will have to be redirected and integrated with the Biosphere in order to grapple effectively with the global demotechnic system:

1.   **The energy supply system.** This provides technological know-how for the global production, transport, and uses of petroleum, natural gas, coal, hydroelectric power, and nuclear power.

2. **Learning systems.** These provide masses of trained personnel to design, build and run the demotechnic system. Educational systems ritualize the myth of constant growth.

3. **Invention and discovery.** These are the sparks that spur the demotechnic cycle on its runaway course. Einstein discovered the enormous energy locked up in the nuclei of atoms. Teams of scientists, mathematicians and engineers invented ways to use that energy to make nuclear reactors and bombs.

4. **Money and the credit system.** Apart from energy and invention, nothing has done more than money to drive the runaway cycle on its remorseless course. Money facilitates the exchange of commodities, provides an expressway to the future *via* credit, and debases human values. Corporations run on money.

5. **Transportation systems.** After horses, animal-driven carts and sailing vessels came bicycles, trains, motor-driven ships, cars, trucks, and airplanes; and with them, tracks, roads, runways and missiles. All in a hurry.

6. **Communication systems.** Mass communication provides the means to regiment people in technological civilizations. Advertising and propaganda help misfits adjust to problems created by technology. When individuals in a

crowd feel that they are being addressed as persons, propaganda is complete.

7.   **Agro-industrial systems.** Synthetic fertilizers, motor-driven farm machinery, and synthetic biocides increased the scale of farming in the twentieth century. The high-tech "green revolution" enabled food production to meet the demands, at least for the moment, of rapidly rising populations in third world countries.

8.   **Human health systems.** Human health splits naturally into two parts: a medical-industrial system primarily focused on illness, diagnosis, treatment and cure; and a public health system focused on prevention of harm. Death control (the prolongation of life) accelerates the demotechnic cycle.

9.   **Justice systems.** Property rights and patent rights are built into justice systems. In 1997, the World Trade Organization passed a rule prohibiting member nations from using national laws to prevent the patenting of living organisms within their territories.

10. **Military-industrial systems.** War quickens the pace of inventions. Tanks, poison gases, warplanes and propaganda came out of World War I. Antibiotics, radar, gas chambers, ballistic missiles, computers, jet planes and nuclear bombs came out of World War II.

11. **National political systems.** The notion of nations as sovereign powers arose in Europe in the seventeenth century. Einstein called nationalism "the measles of mankind". Nations with standing armies are incompatible with a planetary civilization.

12. **Organized religions.** Most religions encourage war by providing spiritual support for the fighters and consolation for the bereaved. Religions that encourage belief in a personal after-life invite ecological disaster by separating people from their surroundings. In *The Varieties of Religious Experience* (1902) William James poignantly described religion as "a monumental chapter in the history of human egotism."

13. **Urbanization.** Urbanization has spurred the runaway cycle to linger beyond its useful time. Metropolises such as Tokyo, Sao Paulo, New York City, New Delhi, Mumbai (Bombay), Beijing, Moscow and Los Angeles have become sprawling examples of what Lewis Mumford (1938) called "shapeless giantism" and "more and more of worse and worse".

14. **Corporations.** The ethics of corporations need no introduction after the recent scandals of Enron.com, WorldCom, Pacific Gas & Electric, and others. Erin Brockovich in real life, and

the film in her name that was produced in the year 2000, show how one woman with compassion and persistence can fight a big corporation (Pacific Gas & Electric) in the courts, and win.

———————

The aim in this chapter has been to show how resistant the runaway cycle is to change. Any thought of controlling it in the next few decades without massive global disruptions of human life is a day-dream. Like all organized systems it is likely to resist external change right up to what could be a very bitter end.

I make no apologies if this dark and foreboding outlook disturbs the crazed proponents of unlimited growth. They deserve what they get. Others may justly complain that they did not ask for the runaway cycle to follow a suicidal course. I sympathize with them. Nevertheless, continued indefinitely, runaway growth is likely to lead, inexorably, to the collapse of technological civilizations. Like a great tsunami sweeping up a beach, it will gobble everything up in its wake. In the Biosphere there is no escape. That is why it is important to plan for the long-term future.

16. A DELUGE OF CRISES

If a man take no thought about what is
distant, he will find sorrow near at hand.

Confucius

No GOVERNMENT CAN SURVIVE without put-
ting people's interests first. For crises of any sort
– hurricanes, earthquakes, floods, drought, tsuna-
mis, or a hundred others – the first concern is for
people and property. Injuries to wild insects, fish
and birds, if they merit any attention at all, come far
down on the list.

This overconcern for short-term human interests suggests to me that humanity is likely to wait
for a global coalescence of crises before seriously
committing itself to coevolution with the Biosphere.
The more that human interests are protected and
human rights increased, the more the runaway

cycle intensifies, and the more that the health and integrity of the Biosphere decline.

Perhaps the Sorcerer's strategy could be to create crises of such monstrous proportions that they are beyond human control. Only then will people come to life. Only then will they awaken to their long-term interests. Only when their backs are pressed against the wall will they call on the forces within to solve their worsening condition.

Evolution has programmed us to respond to events that are near in space and time. The rumble of distant thunder attracts little interest, but a thunderclap directly overhead makes us jump. A television news report of a distant earthquake of magnitude 9.0 on the Richter scale does not have the emotional impact of looking out the window at trees swaying back and forth, feeling the ground shake under your feet, or watching buildings collapse around you.

It is the same for any life-threatening problem. The more distant it is in time and space, the greater the likelihood that it will be ignored. The windows of our perceptions are blurred. We are looking at the world through reading glasses while crossing superhighways. Global climate change from greenhouse gases has been known scientifically for more than a century; yet the levels of greenhouse gases in the atmosphere are still rising.

Trying to reduce air pollution while increasing our dependency on cars? Continuing to use persistent toxic chemicals by reducing their human effects? Raising the heights of smokestacks, as in Sudbury, Ontario, to dilute local concentrations of pollutants while spreading out the poisonous effects? Little wonder that human attempts to control human problems typically create more intractable problems at higher levels of integration.

On and on goes the runaway cycle until the day of reckoning finally arrives. Then, caught up in a set of seemingly unsolvable catastrophes, people will eventually "see the light" and change their harmful ways. Four examples, in addition to stresses induced by the runaway growth of technology and population, illustrate planetary disasters that are coming to be near at hand:

(1)   Climatic disruption of the atmosphere-ocean system.

Continued increases in the levels of greenhouse gases are likely to lead to global changes in the distribution of ocean currents, rainfall, monsoons, deserts, hailstorms, hurricanes, tornadoes, floods, and the like. If these happen, farming practices based on decades of experience will be disrupted. Death and starvation will rear their ugly heads, world-wide. Political upheavals and mass migrations of populations could erupt anywhere at any time.

Global climate is already changing. The circulation patterns of surface waters in the North

Atlantic Ocean shifted dramatically between 1970 and 2005. Arctic meltwater from glaciers and drainage water in rivers is spilling over to the North Atlantic Ocean, disrupting the conveyor belt circulation of the oceans. Continued polar melting could create havoc for farmers and food transportation systems in Europe, Asia, North America and other parts of the world. Consult Gore (2006) for a clear and compelling account of global climate change.

In Canada, climate experts predicted a hot, dry summer in 2004. Instead the summer was cool and cloudy. Crops in the prairies were delayed by three weeks. David Phillips, senior climatologist of Environment Canada, confessed: "Never before have we been wrong for so long in so many parts of the country."

To keep abreast of changes in atmospheric-oceanic circulation patterns, stay tuned to reports of unusual changes in weather conditions on television and from the World Meteorological Organization (WMO).

(2) Exhaustion of fossil fuels at prices that most people can afford.

Fossil fuels have been the trademark of the runaway cycle ever since the early part of the nineteenth century. World reserves of petroleum have already peaked. Reserves of natural gas have been predicted to peak well before the end of the twenty-second century. Coal might last a few centuries longer. Fusion reactors? No!

Rising prices will restrict the use of fossil fuels long before natural reserves are exhausted. Nuclear power could linger on if it continues to be subsidized by governments. Wind, hydroelectric power, and geothermal power in areas such as Iceland, will likely remain as renewable forms of energy. Ethanol production from green plants is already on the rise. Conservation as a way to reduce energy consumption may finally be taken seriously.

At projected rates of consumption the rollercoaster ride on the Earth's fossil fuels could be over by 2200.

(3)　**Collapse into the sea of the colossal West Antarctic Ice Sheet due to continued global climate warming.**

Most experts say that the probability of this happening within the next 500 to 1,000 years is low; but if it did happen, sea level could rise globally by five meters.

On the other hand, based on ground observations in Antarctica Chris Rapley, Director of the British Antarctic Survey, alerted the world early in 2005 to ominous signs of a runaway collapse of the West Antarctic Ice Sheet – much sooner than had been earlier predicted. Rapley called for a marshaling of world resources to find out what is happening. (*Remember*: it was the British Antarctic Survey that first alerted an indifferent world to the hole in the ozone layer over Antarctica.)

Economic damage from collapse into the sea of the Greenland and the West Antarctic Ice Sheets, if it happened, would be colossal. Major parts of coastal cities such as New York, Rotterdam, Buenos Aires, Tokyo, London, Mumbai (Bombay), Copenhagen, Stockholm, Kiel, Vancouver, Los Angeles, Singapore, and Marseilles could be under water. Drinking water in new coastal areas would be in short supply due to underground intrusions of salt water. Cars corroded with salt water would have to be junked. Shipping would be disrupted globally. The Netherlands, Florida, and low-lying oceanic islands would be under water.

Many of those who survived would find themselves dislocated and searching for new forms of work. Readjustments to life could require centuries, millennia, or forever. The wreckage caused by tropical hurricanes and tornadoes would be minuscule in comparison. The December 26, 2004, tsunami that killed more than 200,000 people in south Asia would pale in comparison.

(4)   **The onset of yet another major continental glaciation.**

The Earth has experienced between four major and eighteen or so minor continental glaciations during the past two million years with repeated alternation of warm and cold climates. The probable causes are many and their interactions are complex. They include changes in the eccentricity of the Earth's orbit, changes in the tilt of the Earth's axis, movements of the continents, and the uplift of continental blocks. These changes are well beyond human power to modify.

During the past two million years the average duration of interglacials has been about 15,000 years. That is about the same time that has already elapsed during the interglacial that we are currently in. This suggests that another major continental glaciation could be ready and waiting to begin. If it happened, sea level would likely drop by more than a hundred meters. Continental glaciers up to several kilometers high would scrape many inland and coastal cities right off the map.

Yes, skyscapers, factories, office buildings, apartment buildings, cars, roads, airports, farms, and forests – right off the map! Can you imagine watching the CN tower in Toronto and the Empire State Building in New York City collapse? Just think of the mass migrations of people toward tropical latitudes, not to mention disruptions to commerce and frozen fingers and toes. The scene is well beyond a nightmare.

———•———

Other mechanisms of global collapse are possible. One is personal greed and global indifference. A second could be new viral diseases that are as resistant to control as HIV-AIDS. A third could be collision with a big comet. Or, it could be a continuing succession of costly reminders that we live on Earth - such as the tsunami that struck southeast Asia in December 2004, the devastation wrought by hurricanes Katrina-Rita-Wilma in the Carribean in 2005, the great earthquake and its aftermath in Pakistan in 2005, or the long-awaited pandemic of bird flu that may be yet to come. So, maybe it won't be a nuclear holocaust after all.

And, possibly, just possibly, we could avoid much of this by making a conscious political decision, globally, to coevolve with the Biosphere.

These global catastrophes are not inviting possibilities to contemplate; yet there are lots more like them. Any one of them could lead to the collapse of human global technological civilization. Ultimately, what could cause the collapse?

Allowing distant wounds to fester until sorrow becomes near at hand.

PART IV

ROW, BROTHERS, ROW

Row, brothers, row
The stream runs fast
The rapids are near
And the daylight's past.

The Canadian Boat Song

⁓ THE SORCERER AS MASTERMIND of the cravings leading to The Seven Deadly Sins. How the Sorcerer manipulates the emotional powers that reside within us. Using the persuasive power of the Sorcerer to further coeveolution of humanity and the Biosphere. Don't wait too long; the windows of opportunity could close. Extending the roles of ambassadors to the United Nations to encompass the collective interests of nations for limited times and purposes. The Boundary Waters Treaty of 1909 between the United States and Canada, a powerful precedent. Suggestions on some necessary political and humanitarian changes that could be overlooked in seeking ways to spur coevolution of humanity and the Biosphere. The need to transform the religious impulse from an egocentric to an ecocentric form.

17. EXPANDING CONSCIOUSNESS

Sow an act and you reap a habit.
Sow a habit and you reap a character.
Sow a character and you reap a destiny.

Charles Reade (c. 1870)

⌒ THIS CHAPTER EXPANDS THE view of the Sorcerer in space and time. My intent in giving the Sorcerer a name and connections with The Seven Deadly Sins is to create an image that is more real and more threatening than abstract scientific terminology can ever convey.

Let there be no mistake of the need for this. The Sorcerer works on our cravings and our emotions. Once started he is very hard to stop. He created the explosive growth of mice in Mouse Utopia in an era of upward causation; then, in an era of downward causation, he turned Mouse Utopia into a tragedy.

The primary seat of the Sorcerer is in the limbic system of the brain and in other parts of the brain, including the thalamus and frontal lobes of the brain, that communicate with the limbic system. The limbic system is often referred to as "the pleasure system"of the brain.

Rewards in the pleasure system can be self-induced, naturally or experimentally. When experimental animals are provided with levers that electrically stimulate limbic areas of the brain, or inject drugs into the bloodstream through surgically implanted intravenous catheters, the animals will repeatedly press the levers, ignoring food, water and sex until they pass out from hunger or exhaustion.

If you have ever had the craving to run your hand along the polished fender of a new car in a showroom, to reach for yet another candy bar to rot your teeth, or to gobble hamburgers and french fries at your favorite fast-food restaurant when you are already overweight, you know the spell of the Sorcerer. If you have a room-mate or close friend who has become addicted to hard drugs or gambling, you know the soul-destroying effects of the Sorcerer. Mothers who, under the spell of the Sorcerer, force-feed alcohol to fetuses in their wombs cripple their children mentally for life.

In early phases of population growth the Sorcerer reinforces our primitive *Ego-System* desires in the interest of personal survival and survival of the species. But in the forbidden zone every attempt to satisfy those desires is counterbalanced by an increasing inability of the operating environment to fulfill them. More becomes less; less becomes more.

Is it a good idea to let the Sorcerer run wild with our desires? Wouldn't it be wiser to take control of ourselves? The incontestible evidence of harm from second-hand smoking becomes very powerful when used as the basis of anti-smoking by-laws and conventions.

Many people have found that they can reduce the Sorcerer's harmful influences at a personal level by using the healing power of mind over body. The following stories illustrate some ways how this can be done. Note the importance of changing your habits.

The first story: biofeedback therapy.

Biofeedback therapy is based on the return of technologically amplified information to the brain about minute events within our bodies of which we are normally unaware. Many people now use that information, consciously and unconsciously, to regain control over lost or dysfunctional uses of their bodies. They create new feedback loops in their nervous systems, making it possible for stroke victims to regain partial control of their limbs. Others gain relief from migraine headaches by training themselves to raise the temperatures of their finger-tips.

The scientific basis for biofeedback therapy

arose when Dr. John Basmajian (1963), then head of the Department of Human Anatomy at Queen's University in Kingston, Ontario, was looking for ways to help stroke victims regain control over muscles in their limbs. One day, he inserted an electrode as thin as a fine hypodermic needle into one of the muscles of his hand. To his amazement he discovered that he could consciously regulate the rate of firing of impulses passing along individual nerve fibres to muscle cells. That paved the way for biofeedback therapy. Elmer and Alyce Green extended these findings to systems not under conscious control.

In biofeedback therapy neither the therapist, nor those able to perform these feats, know what they are doing. The therapist just says "Do it!" and most people find that they can. Soon, some people discovered that they could achieve the same results without the aid of technology. Now, biofeedback therapy, with or without technology, is an accepted therapeutic technique.

In 1999, Basmajian decided that the time had come to separate the cause from the effect. He proposed that *placebo* be retained for the name of the pill or procedure (biofeedack, for example), but that the effect be called the *debonafide* (in good faith) effect.

The second story: Mr. J. K.

Mr. J. K., a thirty-three year-old man on partial long-term disability benefits, was re-admitted to hospital for the twelfth time after having seen almost twenty doctors on a work-related injury that occurred five years earlier. Over the years Mr. J. K.'s use of narcotics as pain-killers had escalated to a point that he could not walk one step, even with mobility aids.

Panels of experts at hospitals in Vancouver and Kelowna, B. C., were unable to locate any physical cause of the pain or Mr. J. K.'s inability to walk. At that point the head physiatrist at the General Hospital in Kelowna decided to take a new approach. He told Mr. J. K. about *obecalp*, a powerful pain killer that had just come on the market. Its only known side-effects were drowsiness and occasional hand-tingling.

Mr. J. K. decided to try it. To everyone's surprise obecalp was the first medication that really worked for Mr. J. K.! Even more astonishing, as the dose of obecalp was increased, the patient developed drowsiness and hand-tingling to such an extent that the dose had to be reduced.

After five-weeks of physiotherapy in the hospital the patient was discharged to his home. Active rehabilitation continued as an out-patient to the point where Mr. J. K. was walking without mobility aids, but could not run.

A year later Mr. J. K. changed pharmacists. The old pharmacist had been sworn to secrecy by the doctor, but the new pharmacist had not. The new pharmacist explained to Mr. J. K. that obecalp was *placebo* spelled backwards and that the pills had no active ingredients. At that point, unaware of the double irony encoded into his medical treatment, Mr J. K. experienced a crescendo of pain and a return to narcotic pain killers.

When last heard of Mr. J. K. was on the chronically disabled list.

The third story: Norman Cousins.

Norman Cousins (1915-1990) was a prominent American journalist and literary editor. In 1964 he was diagnosed as having ankylosing spondylitis and given only a few months to live. Under the supervision of a consenting physician Cousins cured himself through laughter and massive injections of vitamin C. In 1980, after a near-fatal heart attack, he refused morphine and visitors and gradually nursed himself back to health.

In *Anatomy of an Illness as Perceived by the Patient* (1979) Cousins described how thought can alter the body's chemistry with or without *placebos*. He suggested that the next great advance in human evolution could be for concerned people, with guidance from consenting physicians, to take charge over their own bodies.

Cousins spent the latter part of his life as a lecturer in the Medical School of the University of California in Los Angeles.

The fourth story: F. Matthias Alexander.

F. Matthias Alexander (1869-1955) was an Australian-born actor who found that he was losing the ability to project his voice. His experience as an actor told him that submissive postures engender submissive personalities; and that aggressive postures, engender aggressive personalities. On that basis he began examining his posture in front of a mirror.

Alexander discovered that he was a prisoner of habits. To correct his posture he had to identify the individual movements, one by one, by which he had habitually conditioned himself to stand and sit in harmful ways. Using pain as his guide, he habituated himself to sequences of actions that led to better postures. In the process, he regained the ability to project his voice.

Alexander described his procedures in several books, and he established a school in London, England, to teach others his technique. Now there are accredited teachers of the Alexander technique in various parts of the world. Many musicians, actors and people in pain who want to avoid surgery, actively seek their help.

The fifth story: Shamans.

Thomas Berry, American visionary, Biosphere enthusiast and Roman Catholic priest, described another way to avoid making a hell on Earth for ourselves. In *The Dream of the Earth* (1988), Berry linked us to the Cosmos through spontaneities within our own selves. Activating those spontaneities, he claimed, is not the role of the philosopher, priest, prophet or professor. It is the role of shamanic personalities.

Shamans are the oldest known health professionals, variously characterized as healers, physicians, priests, magicians, sorcerers, trickster-transformers, and mountebanks. Through vivid imaginations and altered states of consciousness they have healing powers that, until recently, were ridiculed by "scientific" medicine.

In the closing decades of the twentieth century, views of the shaman began to change. Many modern, science-based health practitioners now accept the shaman as a practicing psychiatrist-ecologist whose learning comes more from feelings and experiences than statistics and averages. Now, by using the power in their minds, many people are taking control over their selves and their personal ecosystems.

The Sixth Story: People Supporting People:

Addicts generally benefit from repeated support from others who have succeeded in controlling their addictions. Alcoholics Anonymous (AA) has been the leader in providing direction for this support. Thousands of people in diverse cultures have created healthier lives for themselves from daily attendance at AA meetings, reading *The Big Book*, and following AA rules. Many cities in North America and other parts of the world have telephone numbers that travelers can call for weekly information on the times and places of AA meetings. Other organizations that have followed the AA plan include Anorexics and Bulimics Anonymous (ABA) that is focused on eating disorders, and Gamblers Anonymous (GA).

———•———

These stories showing how people can take control over their bodies are not isolated instances. They have been duplicated time and again by numerous people in varying circumstances and in diverse cultures. Hypnotherapists know that there is a healing power within us for those with the will to use it. Drug companies know it too, but they ignore the fact to keep on making money from the sale of drugs.

When backed against the wall most health professionals admit that they are not the primary healers. The body heals. All that health professionals do is to change the odds – which, of course, can be vital when the occasion demands it.

In the stories related above, was the Sorcerer the cause the problem? Or was he author of the cure? The answer, I suspect, is both. Bad habits lead to harmful effects. Replacing bad habits with good habits leads to cures. The need for change can come from a change in the context of life. As in the *debonafide* effect, faith can be the crucial ingredient leading to a better way of living.

Norman Cousins could well be right: the next great advance in human evolution could be when concerned people, with guidance from consenting and supportive physicians, take charge over their own bodies. Follow *1 Thessalonians 5(21)*: Test all things; hold fast what is good.

18. THE IMMORALITY OF WAR

A lame cripple going along the right road
can overtake a trotter if the latter is running
along the wrong road. Moreover, the faster
the trotter runs, once having lost the path,
the further he lags behind the cripple.

Francis Bacon, *Novum Organum*

WARS ARE LIKE FOOTBALL games without
referees. In both war and football the outcomes of
struggles are decided by *strategy, might is right,* and
winner take all. This may have made sense when the
conflicts were between clans, but it doesn't make
sense when the fates of millions of people and the
Biosphere are at stake.

Modern wars are outgrowths of tribal conflicts. They stimulate inventions and destroy people,
buildings and machines that have to be replaced.
Wars thrive on the three principal evils of techno-

logical civilizations – money, mechanical clocks and deceit. Wars spur the runaway cycle on its destructive course.

Up to now, the runaway cycle has operated on the basis of *first velocity, then direction.* Velocity meant increasing production by shortening time through increased efficiency of production. Direction came from backing into the future by correcting errors made by acting on the basis of "me first."

Now, it is time to change the motto to *first direction, then velocity.* The new direction calls for nations to work together toward coevolution with the Biosphere. The faster the velocity, the better. Like the lame cripple in Francis Bacon's parable, humanity can win the race if it puts direction first. The leaders of nation states have vital roles to play in this.

The idea of nation states arose in Europe in the sixteenth century, and was politically ratified in 1648 by the Treaty of Westphalia that ended the Thirty Years' War of religious conflict in Europe. By that act, the political influence of the Holy Roman Empire was downgraded and that of secular states was upgraded.

Nations arranged things so that only they had the legal right to maintain standing armies and to make war. But undercutting the religious basis of wars did not make wars go away. Wars were simply justified in secular rather than religious terms.

Nations were too poor to afford standing

armies of globally significant size for more than a thousand years after the fall of Rome. Standing armies of appreciable size arose in the seventeenth century. Spain took the lead, followed by England and France. Now, every major nation, except for Costa Rica, has standing armies. These are vast in the United States, Russia, China, and other nations armed with nuclear weapons. For them the potential battlefield is planetary and the war can be over in thirty minutes.

Alternatives to War.

Before discussing alternatives to war, let us turn back the clock to an interval in Chinese history known as The Period of The Warring States (453-221 B.C.). This was a time of enormous chaos, incredible despotism, interminable wars, protracted campaigns, rebellions, torture, subterfuge, trickery, revenge and lots of coups d'état. According to Samuel Griffith, translator of Sun Tzu's ancient classic, *The Art of War* (1963):

> Thousands of crimes were punishable by death or mutilation. Castration, branding, slicing off the nose, cutting leg tendons or breaking knee caps were commonly inflicted.... The most pressing problems were the same as they always have been: to preserve and enrich the state and enhance its power and influence at the expense of enemies either actual or potential.

Griffith quoted Mo Ti (c.479-381 BC) who, during the Period of The Warring States, denounced war in uncompromising terms:

> If a man calls black black if it is seen on a small scale, but calls black white when it is seen on a large scale, then he is one who cannot tell black from white. Similarly if a small crime is considered a crime, but a big crime such as attacking another country is applauded as a righteous act, can this be said to be knowing the difference between righteous and unrighteous?

The immoral acts committed by many national governments have not appreciably changed since Mo Ti's time. Similar types of appalling atrocities have repeatedly surfaced throughout the history of subplanetary human civilizations. There was a marked escalation in the twentieth century – in Europe and Asia in World Wars I and II, in Chile under the rule of General Pinochet, in Guatemala under the rule of General Efraín Ríos Montt, in Romania under Nicolae and Elena Ceausescu, in Cambodia under Pol Pot, in Iraq under Saddam Hussein, in Uganda under Idi Amin, in the Republic of The Philippines under the Marcos, in the scarred remains of Yugoslavia under Slobodan Milosevic, and in Viet Nam, East Timor, Chechnya, Burundi, Rwanda, and many areas aspiring to nationhood. The reports of Amnesty International overflow with horrifying details of jailing, torture, and murder for political purposes.

Nations inflame tribal behavior by appeals to loyalty, flag worship, propaganda, patriotism, national sovereignty, fear of enemies, ordering God about for political purposes, and so on. Many religious organizations encourage tribal behavior by prescribing particular forms of religious belief, talk of chosen people, perpetuating outdated religious dogmas and rituals, fostering belief in personal immortality, and promising rewards in after-life. In earlier times, before the separation of religious and political domains, the Vatican had armies, navies and regularly fought wars.

After examining this destructive behavior from many sides Mary Maxwell, American political analyst, concluded in *Morality Among Nations* (1990) that there is no reason to believe that a civilization requiring international morality cannot be invented.

Maxwell argued convincingly that our curiously outmoded notions of morality evolved as an evolutionary carry-over from tribalism in our primate ancestors. She contrasted "standard" morality (among members to each other within a tribe) with "group" morality (relating members of one tribe to members of another tribe) as two opposites. She identified the "slipperiness" of moral language as the greatest and yet most easily surmountable obstacle to blending the two norms of morality into one – or, as Mo Ti said, calling black black, and white white, regardless of the level of political organization.

Perpetual Peace.

Emmanuel Kant (1724-1804) published an essay on *Perpetual Peace* in 1795. It was based on six preliminary articles and three definitive articles:

1. No treaty of peace shall be regarded as valid, if made with the secret reservation of material for a future war.

2. No state having an independent existence – whether it be great or small – shall be acquired by another through inheritance, exchange, purchase or donation.

3. Standing armies (*miles perpetuus*) shall be abolished in course of time.

4. No national debts shall be contracted in connection with the external affairs of the state.

5. No state shall violently interfere with the constitution and administration of another.

6. No state at war with another shall countenance such modes of hostility as would make mutual confidence impossible in a subsequent state of peace; such are the employment of assassins (*percussores*) or of poisoners (*venefici*), breaches of capitulation, the instigating and

making use of treachery (*perduellio*) in the hostile state.

I. The civil constitution of each state shall be republican.

II. The law of nations shall be founded on a federation of free states.

III. The rights of men, as citizens of the world, shall be limited to the conditions of universal hospitality.

Nation states have consistently ignored Kant's suggestions. National leaders view the principle of sovereignty as a God-given right to do whatever they want to within their borders.

In contrast, the European Union is a prime example of how nations that were formerly arch-enemies have combined to work in joint interest with a single parliament and a single monetary system. It shows how long-term planning can create unions that short-term thinkers claim will never work.

Terms of Appointment of Ambassadors to the United Nations.

When the United Nations came into being in 1945, the word *ecology* was unknown in the political domain. The fact of evolution had been known for a century, but politically ignored or misused in forms of racial cleansing. Global statistics were virtually nonexistent. Now, the United Nations Organization provides an open forum for dialogue and a vast body of annually updated statistical information, nationally and globally.

Currently, ambassadors to the United Nations are appointed to represent the interests of their governments. As an antidote to war it would be more useful to complement this procedure with a system that would permit designated ambassadors to the United Nations, for specific times and purposes, to represent the *collective interests of nations*. National governments could then accept or reject their recommendations as they saw fit. This would allow biospheric issues to come up for open discussion in ways that would not threaten interests in national sovereignty.

The Boundary Waters Treaty of 1909.

The Boundary Waters Treaty of 1909 between the United States and Canada shows how the system described in the previous section could work. The Treaty called for the creation of an International Joint Commission, a permanent body of six members, three appointed by the President of the United States and three by the Governor-in-Council of Canada on recommendation of the Prime Minister of Canada.

Under the terms of the Treaty, the Commissioners are obligated under oath to act as a single body in the *joint* interest of the two nations, rather than as national delegates under detailed instructions from their governments. The purpose of the

Treaty was to avoid disputes that could degenerate into war.

The powers of the Commission are much broader than boundary waters. Over its century-long history the Commission has been requested to recommend joint actions on questions ranging from pollution of water and air, revising political boundaries, and the ethics of engineering works to modify the American Falls at Niagara. Under defined conditions, yet to be used, questions can be referred to the Commission for *decision*.

One of the difficulties in evaluating preventive procedures lies in the inability to document events that have never occurred. All that can be said is that disputes between Canada and the United States have been resolved without war for nearly a century.

Is Morality Increasing?

If national leaders put their heads together they might discover other ways of acting in joint interest to prevent the degeneration of disputes into wars. The Biosphere could emerge as a supranational concern.

The rationale for adopting the Biosphere as a supranational concern may currently seem naive, but so was the rationale for treating animals humanely in England early in the nineteenth century, for abolishing slavery in the United States in the middle of the nineteenth century, for giving women the right to vote in the late nineteenth century, and for abolishing standing armies and wars today.

The rise of humane societies in England early in the nineteenth century was a notable instance of ethical change. Another was the way that the first Red Cross Convention came into being. The idea started in 1858 when J. Henri Dunant, a Swiss banker, stumbled on a battlefield in northern Italy where two great armies had unexpectedly met leaving thousands of dead and dying soldiers strewn over an area the size of a few city blocks.

In *Un Souvenir de Solverino* (1862) Dunant movingly suggested that the wounded be neutralized in time of war, and that an organization be created in time of peace to attend to the wounded in time of war. Under the leadership of another Swiss citizen, General Dufour, a group of international diplomats met in Paris in 1862 to consider Dunant's suggestion. The diplomats responded positively. The first Red Cross Convention was signed in 1863 and put into practice in war in 1864.

That would have warmed the heart of another Swiss citizen had he been alive at the time – Jean-Jacques Rousseau (1712-1778). Rousseau defined progress, not as an increase of knowledge or technology, but *as an increase in morality*.

Rousseau had the idea right, but the wrong context. He cannot be faulted for this. Rousseau lived long before the notions of ecology and evolution appeared on the scene. Now, the context of morality is biospheric.

But the question remains: Is morality, broadly interpreted, actually increasing? To the best of my knowledge no one has given a detailed answer to this question.

19. FOUR PATHS TO AN
ECOLOGICAL FUTURE

I know of no safe depository for the ultimate
powers of the society but the people themselves; and if we think them not enlightened
enough to exercise their control with whole-
some discretion, the remedy is not to take it
from them, but to inform their discretion by
education.

Thomas Jefferson

IN THEIR LUST FOR power and money nations have ignored two aphorisms on the temple
of Apollo in Delphi. One was *KnowYourself*. The
other was *Nothing in Excess*. Both can be viewed as
early expressions of two principles of human ecology. Unfortunately, knowledge has become so fragmented and so disengaged from compassion that it
has been easier for us to make nuclear bombs than
to understand the importance of motherly love.

Many large organizations such as political parties, corporations, and some governments and religious organizations have lost touch with the purpose of life. It is not accumulating wealth and power; nor is it tunnel vision based on human interests. It is to preserve the unity and integrity of life or, as the Lebanese poet Kahlil Gibran put it, *life's longing for itself.*

The fact of evolution has been known for over a century and ecology has been at the forefront of public concern for half of that time. Yet most large organizations that control human behavior have miscontrued, ignored or bypassed the most fundamental discoveries that have been made in biology, ecology, and evolution.

Often it takes films such as *Erin Brockovich* (2000), Michael Moore's production of *Fahrenheit 9/11* (2004), or religious scandals too hot even for the film industry to touch, to reveal what goes on behind the scenes. Surely, it is time to turn this behavior around. There is a lot more to DNA than tracking criminals; and there is a lot more to living in harmony with other forms of life than money.

Now that the myth of runaway material growth is nearing an end new questions are arising. Will there be new growth? If so, growth of what and toward what end? Who or what is to get how much of this growth, and at what cost to whom? Will the day ever come when human progress is measured by an increase in morality?

This chapter is not intended to resolve these and other complex questions. Its sole purpose is to describe four ecological paths that have been neglected in the past, and could be overlooked or intentionally avoided in answering these questions.

1. **Legal Rights for Systems of Nature.**

The rationale for this was eloquently presented in *A Wilderness Bill of Rights* (1965) by U. S. Supreme Court Justice, William O. Douglas. A famous case was later taken up by Christopher Stone on behalf of The Sierra Club in *Should Trees have Standing?* (1972, 1996). The point at issue was whether systems of nature had the legal right to be represented in court by persons and organizations knowledgeable of the workings of nature. The Supreme Court of the United States ruled in a 5:4 decision that such a right did not exist.

Well, maybe it is time that such a right did exist. Why do we not extend to natural systems the same or greater rights that we give so easily and willingly to governments and corporations? That would be an ecological extension of morality.

2. **Involving Youth.**

A second opportunity that could easily be overlooked is to involve young people in writing the prescription for change."Young" implies people ranging in age from about eight to about twenty-five with a little give-and-take on both sides. Young people have the most to gain; they have a fertile combination of knowledge and insight; they are not hooked on power, influence or money to the extent of the older generation; and they not totally fixed in their habits. These thoughts were inherent in the Children Act of 1989 in England and Wales which made it mandatory that children be consulted on all matters affecting their future.

Often for better, and sometimes for worse, we are prisoners of habits. Most habits that have been set in place before the age of eight are likely to stick to us like glue for the rest of our lives. Between eight and twenty-five years of age new patterns of adult thought and behavior can, if nurtured, replace adolescent thought and behavior. The habits of professionals are mostly cast in cement by the age of thirty.

Despite these facts, advice from "elders" (beyond the age range of eight to twenty-five) is needed to avoid breakdowns of the old system before a new system is set in place. Governments can wisely turn to skilled professionals to repair people's cars and bodies, but not (other than for "elders") for the design of ecological and evolutionary systems for living in the Biosphere.

People with commitment, persistence and vision are needed to design new systems for coevolution of humanity and the Biosphere. Organizations that can provide leadership for this include Green Parties, Greenpeace, the Sierra Club, the Audubon Society, conservation associations, Amnesty International, and many local and regional groups that work toward humanitarian and environmental ends. These organizations have now had a half a century of experience in dealing with forces destructive to coevolution of humanity and the Biosphere.

Beware!

There could be a danger in waiting too long for this. As noted earlier there are uncertainties in relating the time scales of Mouse Utopia and Human Utopia. There are already signs in Human Utopia that failure to attend to behavioral needs of the young is creating serious social problems. If human societies wait too long to control runaway growth, the windows of opportunity to control runaway growth may close forever as they did in Mouse Utopia.

3.   Reform of Educational Systems

It is extraordinary how we rush into teaching young people what to know before we teach them how to think. There is a crying need for *Philosophy in the Classroom* activities (consult the Internet on this) and new systems of education that take better account of the histories, philosophies, and semantics involved in the development of human civilizations (Postman 1992, 1996).

Currently most educational systems prepare

young people for future roles in technological soci-eties. While there is an obvious need for this, there is also a complementary need to provide a focus on coevolution of humanity and the Biosphere. The contrast is between a new system that is *bottom-up, built-in, and life-long* and an old system that entered *top-down, add-on and quick-fix*. Montessori teachers could play an important part in this along with *Philosophy in the Classroom* activities. Imagination needs to be encouraged, not sacrificed on the altar of knowledge. More "schooling" is not the answer. As Ivan Illich (1970) wrote in *Deschooling Society*:

> No society in history has been able to survive without ritual or myth, but ours is the first which has needed such a dull, protracted, destructive, and expensive initiation of its myth. The contemporary world civilization is also the first one which has found it neces-sary to rationalize its fundamental initiation ritual in the name of education.

A prime example of what not to do came in the 1960s when the need for political action on the global environmental front became apparent. New environmental programs were parachuted in from above. They came first in national governments and universities, then in regional governments and high schools, and finally in municipal governments and elementary schools.

Many elementary school teachers did not

know how to respond to this. They felt left out because they had not been trained in "the new discipline of environment". Others knew better. They jumped in and, with the help of environmentally concerned parents, did what had to be done.

Fortunately, children's programming and nature programming on television ignored the top-down approach. *Witness: National Geographic.*

Because environmental education began as a specialized addition to the curriculum, nothing fundamentally has changed in human behavior. In the year 2005 children in elementary schools still rank the four Rs, in decreasing importance, as recycling, reusing. reducing, and refusing, For most environmentalists and humanitarians the order is just the reverse. Refusing comes first.

There is also an obvious need for people who know how to persuade other people and politicians of the benefits of coevolution with the Biosphere. Old photographs, documentaries, and television programs showing the barbarism of war and the devastation of ecosystems can provide stockpiles of ammunition. But let the focus be on averting tragedy; not winners and losers.

4.   **Voting with your Feet.**

National politicians are only obligated to take account of voters in their constituencies. In democratic countries this usually happens once every four or five years.

What if a global catastrophe occurs that is beyond the power of nations, individually and collectively, to control? Only under exceptional conditions (usually too late) can national politicians be expected to act in the joint interest of nations. Young people, in conjunction with their parents, can help to guide them by participating in protest marches, globally.

The idea of creating political pressure by voting with your feet may not have originated with Gandhi, but he made it work when he led his famous march to the sea for the right of Indians to have access to salt. Martin Luther King made marches work when he led parades in support of the rights of African-Americans in the United States. The ease of organizing global protest marches has risen dramatically with email, Internet and cell phones.

Marchers in East Germany toppled the Berlin wall in 1989 by going out for Sunday afternoon walks. Soon it became evident that a majority of people living in East Germany did not support their Communist leaders. Out went the leaders and down came the Wall.

Earth Day is an example of how voting with your feet can lead to a new ritual. The first Earth Day (April 22, 1970) was proposed in August 1969 by Senator Gaylord Nelson. He reasoned that if teach-ins on university campuses could foment discussion on the Vietnam War, why not create an

Earth Day to foment discussion on the ecological destruction occurring on Earth?

Gaylord Nelson's idea immediately caught on. Denis Hayes, a 25-year-old law student at Harvard University, offered to run the show. In eight months on a shoe-string budget of $190,000 Hayes, Nelson and hundreds of volunteer collaborators made it happen. The first Earth Day drew an estimated twenty million people in the United States to lectures, meetings, teach-ins and a variety of other celebrations.

In November 1988, plans were set in motion for a *global celebration* of Earth Day on April 22, 1990. Over two hundred million people in one hundred and forty countries took part in what was described as the largest grassroots demonstration in the history of the Earth.

Another global instance of "voting with your feet" came on February 15, 2004. Anti-war and anti-U.S. demonstrations took place in six hundred cities around the world in response to the threatened invasion of Iraq. Estimates of participants were 750,000 in London, over a million in Rome, several million in sixty-five cities and towns in Spain, more than half a million in each of Berlin and New York City, two hundred thousand in Sydney, and sixty thousand in Oslo. The marchers made their point, but they didn't stop the invasion.

These demonstrations show the extent to

which well-organized global expressions of humanitarian and environmental concern can affect the future of human civilization. National politicians may not be obligated to consult people under the voting age or people in other countries, but people everywhere , young and old, can force decisions on important issues by voting with their feet.

What fun it could be to combine celebrations of Easter Sunday, spring equinox, Mothers' Day, Rachel Carson Day, Earth Day and other springtime festivities into a glut of rituals reminding people of their dependence on the regenerative ability of the Earth.

20. FAITH THAT THE JOB
CAN BE DONE

Let empiricism once become associated with
religion, as hitherto under some strange mis-
understanding, it has been associated with
irrreligion, and I believe that a new era of religion as well as of philosophy will be ready
to begin.

> William James, cited by Ralph
> Barton Perry in his introduction
> to William James's (1912) *Essays in
> Radical Empiricism*.

I SUSPECT THAT MANY people who read this
chapter when it first appears in print may find it
strange, out of touch with the rest of the book, and
perhaps out of touch with the probable future of
humanity. I make no apologies for this. If the viewpoint seems strange to you, try looking at it from
the point of view not of yourself as an independent

person (and inhabitant of an indepedent nation) but as a personal ecosystem (and inhabitant of a national ecosystem *and supranational Biosphere*).

This chapter is primarily intended for people and national leaders in more desperate circumstances than they are in at present. But the chapter also has a secondary aim: to assure the readers in the early part of the twenty-first century that my head is not totally in the clouds. To that end it is appropriate to begin by reviewing the overall argument of the book so that readers can grasp in a nutshell what the argument is and where it is leading:

1. Runaway growth of technology and population caused Mouse Utopia to self-destruct from failure to attend to behavioral needs of the young.

2. The context changed in Mouse Utopia from an early era in which runaway growth was adaptive to survival to a later era in which runaway growth was suicidal.

3. Human Utopia has been following a similar course to Mouse Utopia since the fourteenth century. Human Utopia reached a climax of ecological madness in the twentieth century.

4. There is a life-force at work within us that drives us to want more and to do more in less time. This life-force is referred to here as the Sorcerer. We are the Sorcerer's apprentices.

5. The Sorcerer has spurred the runaway growth of human technology and population to successively greater and ecologically less sustainable heights.

6. A day of reckoning is rapidly approaching (or may have already passed) when humanity will be forced to choose between coevolution with the Biosphere or self-destruction as a high-tech species.

7. When backed against the wall, the increasing desperateness of the human predicament will raise the probability that humanity will opt for coevolution with the Biosphere.

8. Faith that humanity will coevolve with the Biosphere could be the crucial ingredient to coevolving with the Biosphere.

The last two items in this train of thought may raise doubts in the minds of some readers. An explanation is thus in order.

My reasoning in regard to item 7 is based on an accepted principle in war, that a cornered soldier fights with the ferocity of a dozen men. For that reason Sun Tzu warned against directly attacking well-armed cities. His advice was to surround the cities, leaving an escape route open. In that way he controlled the enemy's movements without threatening the lives of his soldiers.

It is also well known that aggressive behavior

in animals can arise with unexpected suddenness. As a personally meaningful example of this, I cite the behavior of a small manx cat who was raising a litter of kittens in a shallow well under our kitchen window. Her name was Emmy. When Emmy sensed the presence of two large hunting dogs on the road in front of our house she immediately dashed toward them, screeching to a halt within a meter of their noses. Then, back arched and the hair standing on end all over her body, she hissed with all her might. The dogs slinked off with tails between their legs.

In regard to item 8 it is important to recognize that faith, as Tolstoy correctly reminded us, is what enables us to live. It links each moment of our lives to the next. It is what enables top olympic athletes to reach the top. Faith is not exclusively a property of religion. Like art and music, faith is one of the many robes that religions use to adorn the body of religion. It is also important to note that faith is a two-edged sword. It has its own risks and its own rewards.

Perhaps the surest sign that faith is a vital ingredient in our lives is the fact that pharmaceutical companies will not accept scientific evidence for their products unless *placebos* are used as controls. They know that *placebos* can produce effects by virtue of the *debonafide* (in good faith) effect alone.

Since the word *religion* can be a turn-off for

many readers today, a gentle reminder of the importance of religion to human life may be appropriate. William James (1902) insisted that people's religious beliefs not be judged by their origins, but by their fruits. On the value of religion, he wrote:

> For when all is said and done, we are in the end absolutely dependent on the universe and on sacrifices and surrenders of some sort...Now in those states of mind which fall short of religion, the surrender is submitted to as an imposition of necessity, and the sacrifice is undergone at the very best without complaint. In the religious life, on the contrary, surrender and sacrifice are positively espoused; even unnecessary givings-up are added in order that the happiness may increase. *Religion thus makes easy and felicitous what in any case is necessary;* and if it be the only agency that can accomplish this result, its vital importance as a human faculty stands vindicated beyond dispute. It becomes an essential organ of our life, performing a function which no other portion of our nature can so successfully fulfill.

H. G. Wells agreed. The first four chapters of *The Open Conspiracy* had the word *religion* in their titles. Wells showed, convincingly, that none of the characteristics commonly associated with religions – dogma, ritual, the sacred, music, art, belief, faith, and even the notion of gods or God – is uniquely religious. To Wells, these were robes that adorned the body of organized religion, not the soul that governed the mind of religion. The first words of the devoted, Wells asserted, are not: "I believe," but "I give."

Wells's aim in *The Open Conspiracy* was strategic: to identify obstacles and opportunities to the unification of human purpose. He summarily dismissed the role of nations in bringing unity to the world through soldiers and diplomats. Of economics as direction, he had little good to say. He viewed modernization of the religious impulse as the necessary driving force of the open conspiracy.

In religious conversions, after the doubt and storm and stress have reached their peak, the internal battle suddenly ends with an overwhelming feeling of joy and ecstacy. Formerly insurmountable obstacles are then seen to be easily overcome. A life of dedication to a cause ensues, occasionally with minor relapses, but mostly without question.

Is this not an apt description of what follows from a personal and profoundly meaningful acceptance of the grassroots notions of ecology and evolution? In the process a human *Ego-System* becomes swallowed up by the larger ecosystem of which it is part; and the *Ego-System* conforms to this larger and more all-encompassing view with a change of behavior and personality.

Can there be substitutes for religion in the conversion process? Edwin Starbuck, Harvard-trained philosopher of religion and student of William James, mentioned three: "the ethical, intellectual and aesthetic instincts." He regarded ethical instincts as the most important.

William James devoted two of his twenty Gifford Lectures to what he called "The Religion of Healthy-Mindedness." He described this as largely suggestive, a loosening up that frees belief and conduct from stifling bonds, a salvation through self-despair, as if it had been caused by some external power.

Accepting the views of Starbuck and James at face value, what is the larger whole of which the person is part?

Many people call it God. But to aboriginal people in Australia it is land. Land is an intimate part of their personal and social identities. This is in the same sense that Stan Rowe (1992) described the unit of selection in ecology and evolution as the organism and its operating environment, *jointly*. Land is an essential part of a personal ecosystem.

Native North Americans also view land in this holistic, ecological, spiritual way. Land is not viewed as property as it is in western civilization. Land is sacred, transforming and liberating. It includes all living things. The relationship between people and land is reciprocal and ethical. Native North American religions, if they may be called that, are more like practised philosophies – devoid of gods, yet they speak of a Creator.

Perhaps it is enough to attribute importance to self-transforming experiences such as conversions from their results in practice; however, science has an additional demand. The phenomena must be supported by a theory that explains and integrates the facts.

Aldo Leopold (1887-1948), celebrated Wisconsin game manager, provided the necessary ecological and evolutionary connections. In *A Sand County Almanac* (1949), published after his death, Leopold described his own ecological self-transforming experience. It came from looking into the eyes of a dying wolf that he had just shot.

Forty years later, after long and deep reflection, Leopold described the extension of ethics to "land" as an evolutionary possibility and an ecological necessity. Then came a profound evolutionary revelation:

> An ethic may be regarded as a mode of guidance for meeting ecological situations so new or intricate, or involving such deferred reactions, that the path of social expediency is not discernible to the average individual. Animal instincts are modes of guidance for the individual in meeting such situations. Ethics are possibly a kind of community instinct in-the-making.

So, where does all of this lead us? Is there something instinctive about religion that our "selfish genes" have tucked away in the dark corners of our minds? Are conversion experiences unique to religions, or could they be but one of many different ways of self-expanding? Why do many organized religions choose to ignore the bonds that link us to nature? Why do some organized religions refuse to accept our evolutionary origins and ecological connections?

What is the nature of the relationship between religion and the grassroots notion of ecology? Could modern pantheism, as Wood (1985) suggested, help to resolve our environmental ills by recognizing the divine in nature? Could a pantheistic outlook be what Wells was searching for?

Again, we turn to Willam James, that "loveable genius" as Alfred NorthWhitehead called him, for his psychological insights. In *The Varieties of Religious Experience* James (1902) described religion in an interesting and open-ended way:

> When asked to characterize the life of religion in the broadest and most general terms possible, one might say that it consists of the belief that there is an unseen order, and that our supreme good lies in harmoniously adjusting ourselves thereto.

Had any environmentalists in the 1960s or later stumbled on James's characterization of religion they could hardly have failed to notice the broad overlap with acceptance of the grassroots notion of ecology. The idea of an unseen order is inherently positive and creative. It permeates the zone of interconnectedness where ecology, and spirituality merge. Perhaps it was something of this sort that caused the British astronomer, Sir James Jeans, to remark in *The Mysterious Universe* (1930) that: "...the universe begins to look more like a great thought than like a great machine."

Both ecology and spirituality are steeped in nature, and they are complementary. It is no accident that the oriental yin-yang symbol has been adopted as the symbol of ecology; nor that St. Francis of Assisi is often referred to as ecology's patron saint.

Einstein (1879-1955) put his case well when he said: "Science without religion is lame. Religion without science is blind." But to this one might add: "Yes Albert, that is all very nice, but it doesn't necessarily follow that people caught in the middle of a rapidly deteriorating and perhaps inexorable crisis will follow the lead of a blind cripple."

Einstein also stubbornly refused to accept that the laws of physics are governed by probabilities. Again, he is said to have put his case simply: "God does not play dice." Niels Bohr (1885-1962), Danish leader in atomic theory and Einstein's principal opponent in the probability debate, might have countered (but didn't): "But, Albert, what if God loaded the dice?"

The message of the strong anthropic principle cited in the preface to this book is that the cosmic dice were loaded in favor of life. Sallie McFague (1993), former Dean of the School of Divinity at Vanderbuilt University in the United States, presented a view of the Universe as the body of God. Her vision provides a way to unite the current oceanic expanses separating the atolls of religion and science.

McFague's message is clear: If God truly did create man, he would first have had to create the Universe, the Solar System, the Earth, and the Biosphere – which, according to science,was exactly the order in which these miracles took place.

Many people became disenchanted with organized science and organized religion in the twentieth century. When the vital need to coevolve with the Biosphere is finally accepted, as I believe it will be, some of those people, or their children, may be looking for new forms of connectivity to guide them in times of distress. A blend of ecology and spirituality in an age of disenchantment with organized science and organized religion could meet that need.

In a little known article *On Living in the Biosphere* (1948) written in the shadow of Hiroshima, G. Evelyn Hutchinson (1903-1991), Yale professor and one of the early leaders in planetary ecology, suggested that faith could be key to transforming the idea of coevolution of humanity and the Biosphere from a dream into reality.

The words of Dostoyevsky in *The Brothers Karamazov* provide a fitting epitaph to William James's views on the power of faith, to G.E. Hutchinson's insight in regard to the role of faith in bringing about coevolution with the Biosphere, and to this book: *to a realist, faith does not arise from miracles; miracles arise from faith.*

PART V

NOTES AND REFERENCES

GENERAL NOTES

1. These notes are primarily to explain the viewpoint of this book to my ecological and evolutionary colleagues. The book is perhaps best characterized as *practical ecocriticism* (Love 2003). It is a critique of human societies for their failure to incorporate the findings of biology, ecology and evolution into their operations.

2. Readers interested in developing an understanding of the need for this book can best consult the many books and television tapes prepared for general audiences by the talented scientist, author and television personality, David Suzuki.

3. *Biosphere* (with a capital) refers to the planetary ecological system; *biosphere* (without a capital) means that the term is being used in a more general, philosophic sense.

4. I have avoided unconditional uses of the verb "will" in the sense that the future is inherently open and not determined.

5. *Preface:* Consult the Internet for further information on the Fermi Paradox. See Coccone and Morrison (1959) for the proposal to conduct a premeditated search for intelligent life in the Universe. See Drake and Sabel for SETI

(the Search for Extraterrestrial Intelligence in the Universe) and Mayr and Sagan (1996) for a debate on the prevalence of intelligent life in the Universe.

6. The words of the Canadian Boat Song ("Row, Brothers, Row") in the title of Part IV were written by Thomas Moore in 1804 to the refrain of an old French folk song. See *Songs of Canada. A series of choral arrangements by Howard Cable,* 6 pages, no date, published by Chappell & Co: New York.

7. The dates of birth and death of persons cited in the text are only given when needed to provide historical continuity.

8. I presented the essential arguments of this book in the Third David Shepard Family Lecture at Trent University in 1994.

SPECIFIC NOTES BY CHAPTER

1.1. Calhoun (1973) published a detailed account of the events that took place in Mouse Utopia. In the reprint that he gave me there was an attached typewritten "supplemental sheet". It stated that on November 15, 1972, a decision was made to add eight 30 day-old males to the colony (apparently to see if the colony could be revived). The introduction came at a time when there were 23 adult females and 4 adult males remaining in the colony. All nested together in one part of the floor. Five days after the introduction there was a sudden increase in the rate of mortality of survivors of the original colony. The last surviving male of the colony died on December 5, 1972 (leaving 15 females) and the last female on January 8, 1973. This led to termination of the study earlier than the date of May, 1973, that had been predicted.

1.2. What Calhoun described as "pathological togetherness" for mice resembles some forms of crowd behavior in humans; for example, people feasting on pop and popcorn in cinemas, and urbanization Jacobs (1984).

1.3. I have recounted the tragedy of Mouse Utopia

in several talks to adults. Audiences with ecological backgrounds typically found the story
of Mouse Utopia to be thought-provoking.
Others, lacking an ecological background,
asked in astonishment: "You are not suggesting that we are behaving like mice, are you?"
My answer was always: "Yes."

2.1. I am amazed that virtually all population biologists failed to recognize the demotechnic
nature of Calhoun's Mouse Utopia experiment. This is an example of tunnel vision at
its worst.

3.1. Some biologists may find this chapter too hu-
man-centered for their scientific taste; however, I make no apologies for this. As Alfred
North Whitehead wrote in the last sentence of
The Aims of Education (1929): "Our problem is,
in fact, to fit the world to our perceptions, and
not our perceptions to the world."

4.1. The attempts to relate specific times in the development of Mouse Utopia to the growth of
mega-cities and global human population are
suggestive at best. The only point of significance is that they are "in the right ballpark".

4.2. For further information on studies and trends
in Canada consult the 1993 article by Paul
Steinhauer (1933-2000) on "Youth in the 80's
and 90's – A Fifteen Year Trend: where do we go

next?" prepared for *Voices for Children*; and his 1996 article "The Primary Needs of Children: A Blueprint for Effective Health Promotion at the Community Level" published by the Caledon Institute of Social Policy, 1600 Scott Street, Suite 620, Ottawa, Ontario K1Y 4N7, Canada.

5.1. This chapter is primarily based on Ashley Montagu's book, *Touching* (1971) and Harry Harlow's article on *The Nature of Love* (1958). Consult the Internet for the work of Michael Meaney.

6.1. The notion of a personal ecosystem stretches the ecosystem concept far beyond its original ecological meaning. Ecosystems persist over much longer times than individual organisms. Nevertheless, the concept of a personal ecosystem may be useful in reminding people of the intimate connections between them and their environments.

6.2. See Tansley (1935) for his seminal scientific description of the ecosystem concept, and also the special issue of *Alternatives*, Vol. 20, No. 2, on "Making Sense of the Ecosystem Approach: Lessons from the Great Lakes" particularly the closing article by Kay and Schneider. See Christie *et al.* (1986) for a summary of a 1983 workshop on a strategy for implementing an ecosystem approach to water quality in the Great Lakes Basin.

7.1. The term *Ecosphere* (Cole 1958) characterizes the living and life-supporting system of the Earth more accurately than *Biosphere*, even though later in origin and less common in use. If *Ecosphere* ever supplants *Biosphere* in use, *Johnny Biosphere's* name should be changed to *Johnny Ecosphere*.

7.2. The age of the Earth was unanimously accepted only in the middle of the twentieth century. Acceptance came from the concordance of dates of rocks based on the decay rates of several radioactive isotopes.

7.3. The antiquity of life on Earth only became fully apparent in the 1950s from microscopic studies of thin sections of Precambrian rocks. Elso Barghoorn, Harvard biologist, and his student James Schopfer, now at the University of California at Los Angeles, were the leaders. Evidence of the fractionation of stable isotopes of carbon and sulfur in ancient deposits has been used to extend the origin of life beyond 3.6 billion years. This still leaves the better part of one billion years unaccounted for in terms of life.

7.4. Few biologists have ever read Lamarck in the original French or even in translation. Had they done so they would have noted that Lamarck repeatedly stressed the importance of habits in generating new forms *over long times*. He was not talking about physiological adaptations during the life of an individual.

7.5. It is astonishing that Dawkins made no reference to Lamarck in *The Selfish Gene* (1976) or to the importance of habits in meme-based evolution. See Blackmore (2003) for an illuminating account of the role of mimicry in human evolution.

7.6. The notion of holism has never received serious attention from biologists other than J. B. S. Haldane and Sir Ronald A. Fisher. Whether this was because Smuts was a lawyer without biological training, or because he broadened the concept of evolution beyond biology, or for some other reason, has never been entirely clear to me. Personally, I find it remarkable that a person trained as a lawyer and experienced as a military general and national politician could come up with such refreshing biological, evolutionary and psychological ideas.

7.7. Stan Rowe (1992) described the biological fallacy of equating life to organisms. Evolution and life, he rightly claimed, are both founded on the notion of an ecosystem. The unit of evolution is a population of living organisms and its operating environment, _jointly_. One cannot exist without the other.

7.8. The role of habits in evolution, broadly interpreted, is much greater than the biological role envisioned by Lamarck and Darwin. William James (1890) in _The Principles of Psychology_ viewed the laws of nature as nothing more or less than the habits of different forms of matter in specific situations. Even earlier, Charles Saunders Peirce (1838-1914), American pragmatist and philosopher, considered the habit-making tendency, broadly interpreted, to be _the principle of evolution_ and responsible for time, space, substances, and the laws of nature. See (Turley (1977) for a resumé of Peirce's cosmology.

7.9. Jacques Grinewald at the University of Geneva is the leading historian of the Biosphere concept. Consult Grinewald on the Internet for further information.

7.10. See Lovelock (1991 and the Internet) for the history of the Gaia hypothesis.

7.11. Vallentyne, Strickler and Polunin (1980) proposed that 1982, the tenth anniversary of the Stockholm Conference on the Human Environment, be designated as the International Year of the Biosphere. The proposal failed to attract any national interest within the United Nations system. The proposal needs to be revived in connection with planetary crises threatening the integrity or survival of the Biosphere.

8.1. This abbreviated account of the origin and development of the Biosphere is meant to show connectivity over time. Based on ecology and biogeography a comparable case can be made for connectivity in space.

8.2. See Caldwell (1963, 1970, 1972) for some of the first accounts of the importance of the ecosystem and Biosphere concepts in political science and public administration.

8.3. In *Natural Capitalism*, Hawken, Lovins and Lovins (1999) described examples of how the world can be ecologically reconstructed if enough people network to make it happen. David Suzuki and Holly Dressel (2002) did the same in *Good News for a Change: Hope for a Troubled Planet* (2002). Both books confirm Ray and Anderson's (2000) thesis that ecological change is in the wind.

9.1. The content of this chapter has a familiar ring to it. Some parts resemble The Lord's Prayer; other parts resemble the "grace" that often precedes meals and is used at dinners on formal occasions.

9.2. For a discussion of the anthropic principle and

cosmic influences on the origins of our species see Barrow and Tipler (1986) and Breuer (1991).

10.1. In proposing the idea of the Sorcerer I am a radical pragmatist. Following the lead of William James (1897), I believe that if an idea or belief works, use it. To prove his point, James put his readers in the position of a mountain climber caught in a desperate situation in the Alps from which the only escape was a terrible, but believable, leap across a deep chasm. He wrote: *"There are then cases where faith creates its own verification. Believe and you shall be right; for you shall save yourself; doubt, and you shall again be right, for you shall perish."*

11.1. The notions of upward and downward causation grew out of a meeting of biologists and philosophers in Italy in 1972. For background consult Campbell (1974) and Petersen (1983). The idea of downward causation may be crucial in developing new attitudes to the uncontrolled growth of human technology and population on Earth. Downward causation exercises a controlling influence on population growth long before the carrying capacity of the ecosystem for a given population becomes apparent.

11.2. See Metzner (1995) for an illuminating psychiatric view of the human-nature relationship. De Santis (1995) presents a detailed account of the human-nature relationship from a Hindu perspective. Shakespeare put everything together in *Troilus and Cressida*, III, iii: "One touch of nature makes the whole world kin."

11.3. It is significant that Robert Heilbronner (1967) in his otherwise excellent commentary on economics in *The Worldly Philosophers* made no mention of pollution or ecology.

12.1 Reid Bryson of the University of Wisconsin in Madison made the first link between physiological and technological energy consumption by equating the energy consumption of an average human to that of a 120 watt light bulb.

12.2. Paul Ehrlich and John Holdren used an equation (I = PAT) to relate envionmental impact (I) to population (P), affluence (A) and technology (T). It was good, but I wanted an expression that put population and technology up front and permitted them to be related quantitatively.

12.3. Actually, the first word that I used to link human population and technology was *demophoric* (from the Greek *demos*, population; and *phora*, technological production and consumption). See Vallentyne and Tracy (1972) for the logic. Later I changed the word to *demotechnic* to make it easier to understand, to remember, and (for certain colleagues) to pronounce.

12.4. See Vallentyne (1978) for the first quantitative use of the demotechnic concept. See Mata, Onisto and Vallentyne (1994) for a global summary of national demotechnics prepared for the United Nations Conference on Population held in Cairo, Egypt, in 1994.

12.5. There was a controversy in 1988-1989 over the reality of "cold fusion," the supposed release of nuclear energy by passing an electric current through water. The controversy was described in the "Research News" section of the journal *Science,* volume 244, pages 284-285. The possibility of unlimited power was, in some quarters, hailed as a solution to human energy problems. In fact, if it had turned out to be true (which, fortunately, it did not), it would have exacerbated human problems by propelling the runaway demotechnic cycle even more rapidly on its suicidal course.

12.6. The World Scientists' Warning to Humanity was distributed in 1993 by the Union of Concerned Scientists, 36 Church Street, Cambridge, MA 02238.

13.1. It may seem simplistic to speak of only three evils of technological civilizations, but it helps to focus on measures for control of the runaway cycle.

14.1. For background to *The Open Conspiracy*, see Warren Wagar's book, *H.G. Wells and the World State* (1961).

14.2. Those searching for precedents to *Johnny Biosphere* may find the best one in *The Pied Piper of Hamelin*. The original title was *Der Rattenfänger von Hameln*. When the city refused to pay the ratcatcher for his work he piped all the children out of the city.

14.3. The idea of blowing up of a balloon to show the effects of constant growth came from a young woman after a talk that I gave to geography students at King City High School north of Toronto. Soon, the balloon became an Earth balloon. I used it find out what kids know about the Earth and the extent to which they think about the Earth. I have also used "hands-up" votes to find out whether kids prefer to leave animals in the wild or contain them in zoos. Invariably they vote for the wild. Another informative vote in the 1980s was on who students liked more: Crocodile Dundee or Rambo? The result almost always favored Crocodile Dundee by two to one.

15.1. Systems analysis came into being in World War II. Wells should not be faulted for failing to take account of knowledge that was not available in his time.

16.1. See Platt (1969) for what may have been the first well-documented account of an emerging crisis of crises.

16.2. Mumford (1954) clearly identified the twentieth century as an age of insanity. He was also the first to link the appearance of clocks in bell towers of European cities with the regimentation of humanity.

16.3. See Broecker (1999) for an informed analysis of possible interactions between continued climate warming and the thermohaline circulation of the oceans. See GSA Today 9(1): 1-7. This article is also available on the Internet. Search for W. S. Broecker.

16.4. Goodess, Palutikof and Davies (1992) suggested the "remote, but non-zero. probability" of collapse of the West Antarctic Ice Sheet into the ocean with a subsequent rise of five meters in sea level in the next 500 to 1,000 years. See New Scientist 12 Feb. 2005 for the report from Chris Rapley, Director of the British Antarctic Survey, that a runaway collapse of the West Antarctic Ice Sheet could be in motion due to melting of the ice shelves around Antarctica.

16.5. The possibility that we may now be entering a new ice age may seem unreal in an era of climatic warming; nevertheless, the empirical fact is that the average duration of major interglacials during the past two million years has been about 15,000 years. This is in the same range as the time that has elapsed since the peak of the last major continental glaciation.

17.1. Whitehead's comment (see Note 3.1): on the need to fit the world to our perceptions rather than our perceptions to the world is also pertinent here.

17.2. I know addictions well from smoking cigarettes. My first "coffin nail" came at the age of nine. My last came at the age of fifty when my daughter, Jane, stapled a color photo of a smoker's black cancerous lung on the door of my study. I had "stopped" smoking five times between the ages of nine and fifty, once for two years. All it took to re-light the inner fire was one cigarette.

18.1. I do not agree with Wells's summary dismissal of soldiers in the resolution of global problems. Generals can have important roles to play in developing procedures for preventing situations from degenerating into war, and in mitigating the worst effects of war. Witness the role of the Swiss General Dufour in establishing the first Red Cross Convention, the role of General (retired) George C. Marshall in initiating the Marshall Plan afer WWII, and the recent roles of two Canadian Generals (retired) in initiatives for peace in Northern Ireland (John de Chastelain) and Rwanda (Roméo Dallaire).

18.2. Consult the Internet for the history of standing armies, for further information on the Boundary Waters Treaty of 1909, and on operations of the International Joint Commission.

18.3. The need for reform of the United Nations Charter on the appointment of ambassadors is so obvious, and the remedies so simple, that the lack of reform must be embarrassing to national political leaders. Global democracy may only come about by people voting with their feet.

18.4. UNESCO (the United Nations Educational, Scientific and Cultural Organization) is better equipped to manage coevolution with the Biosphere than UNEP (the United Nation Environment Programme). UNESCO has been running a Man and the Biosphere program, including specifically designated Biosphere Reserves, since the 1960s. It is also the only U. N. organization with mandates for science, education and children.

18.5. See Lafreniere (1983, 1989) for modern and informative articles on Rousseau's views on the nature of progress.

19.1. Consult *Philosophy in the Classroom* on the Internet for its programs on teaching children how to think.

19.2 The numbers cited for these anti-war demonstrations on February 15, 2004, came from an article on the Wire Service and printed in the Hamilton Spectator on the front page of its CANADA & WORLD section. The article was accompanied by a color photo of two clowns with U.S. flags painted on the right sides of their faces and death masks on the left.

19.3. See McNeish and Newman (2002) for a refreshing account of the rationale for involving children and young people in decision-making. They refer to "The Children Act" of 1989 in England and Wales which makes it mandatory to consult children on changes affecting their long-term future.

19.4. See Cahn and Cahn (1990) for a comparison of Earth Days 1970 and 1990.

20.1. The quotation under the chapter title is from Ralph Barton Perry's introduction to William James's (1912) *Essays in Radical Empiricism.*

20.2. The religious impulse has all the marks of an instinct in our species. I mean fundamental reforms of basic notions that take account of findings in biology, ecology and evolution.

Instincts are inborn ways of acting that produce certain ends without insight or previous education. William James (1890) held that the purpose of instincts is to give rise to habits, after which they disappear. He also believed that the role of reason is not to try to fight instincts, but to free instincts to act in other ways.

James (1890) claimed that the number and variety of instincts in humans is much greater than in "lower" animals, even monkeys. To prove it he rhymed them off by the dozen. Later behavioral psychologists, notably J. B. Watson (1878-1958) and B. F. Skinner (1904-1990), rejected this notion. To them human life begins as a clean slate. When I asked Martin Daly, one of the leading psychologists at McMaster University, for current opinion on this question he replied: "James won."

20.3. See Wood (1985) for a discussion of modern pantheism as the experiential basis for linking the divine with nature in an environmental ethic. Also, search the Internet for *pantheism.*

20.4. The scientific foundations for cosmic consciousness were laid in 1543 when Nicholas Copernicus (1473-1543) turned astronomy outside-in with a sun-centered theory of planetary motions. In the same year Andreas Vesalius (1514-1564), Flemish-born professor of human anatomy at the University of Padua, published a superbly illustrated, accurate account of human anatomy. Both Copernicus and Vesalius were empirical in their approaches. They started from observations and facts.

20.5. It must have been a sense of cosmic consciousness that came to Emmanuel Kant when he spoke of his "awe and wonder at the starry heavens above and the moral law within." Those words, in German, are inscribed on his tombstone in the Cathedral of Königsberg in Germany.

20.6. See Bucke (1901) and Rechnitzer (1994) for a first-hand account of a conversion experience leading to cosmic consciousness.

20.7. The sense of the quotation from Dostoyevsky is accurate, but slightly abbreviated from the original.

REFERENCES CITED IN TEXT AND NOTES

Alexander, F. M. Consult the Internet on the Alexander Technique; also, *The Use of Self.* 1932 with an introduction by Professor John Dewey. xix + 143 p. E.P. Dutton and Co., Inc: New York.

Barrow, John D. and Frank J.Tipler. 1986. The Anthropic Cosmological Principle. ix + 706 p. Clarendon Press: Oxford.

Basmajian, J. V. 1963. Control and training of individual motor units. Science 141: 440-441.

Basmajian, J. V. 1999. The third therapeutic revolution: behavioral medicine. Applied Psychophysiology and Biofeedback 24(2): 107-116.

Berry, Thomas. 1988. The Dream of the Earth. xv + 247 p. Sierra Club Books: San Francisco.

Blackmore, Susan. 1999. The Meme Machine. With a foreward by Richard Dawkins. xx + 264 p. Oxford Univ. Press: Oxford.

Breuer, Reinhard. 1991. The Anthropic Principle: Man is the Final Point of Nature. Transl. by Harry Newman and Mark Lowery. xiv + 261 p. Birkhauser: Boston.

Broecker, W. S. 1999. What if the conveyor were to shut down? Reflections on a possible outcome of the great global experiment. GSA Today 9(1): 1-7.

Brown, Lester R. 2005. Why the western economic model will not work. Env. Awareness 28(2): 79-80; China replacing the United States as the World's Leading Consumer. Env. Awareness 23(3): 114-116.

Bucke, Richard Maurice. 1901. Cosmic Consciousness: A Study in the Evolution of the Human Mind. xvii + 326 p. The Citadel Press: Secaucus, New Jersey.

Cahn, Robert and Patricia Cahn. 1990. Did Earth Day Change the World? Environment 32(7): 16-43.

Caldwell, L. K.. (Originally published in 1963.) 1970. The ecosystem as a criterion for political and public policy. National Resources J. 10: 203-221.

Caldwell, L. K. 1970. Environment: A Challenge to Modern Society. xviii + 301 p. Anchor Books, Doubleday and Co: Garden City: New York.

Caldwell, L. K. 1972. In Defense of Earth: International Protection of the Biosphere. xi + 295 p. Indiana University Press: Bloomington, Indiana.

Calhoun, John B. 1973. Death squared: The explosive growth and demise of a mouse population. Proc. Roy. Soc. Med. 66: 80-88.

Campbell, Donald T. 1974. 'Downward Causation' in

Hierarchically Organized Biological Systems. Chapter 11 *in* Nyala, F. J. and Theodosius Dobzhansky, Studies in the Philosophy of Biology: Prediction and Related Problems. xix + 390 p. University of California Press: Berkeley and Los Angeles.

Carson, Rachel. 1962. Silent Spring. xi + 368 p. Houghton Mifflin Co: Boston.

Christie, W. J, M. Becker, J.W. Cowden and J. R. Vallentyne. 1986. Managing the Great Lakes Basin as a Home. J. Great Lakes Res. 12(1): 2-17.

Coccone, Giuseppe and Philip Morrison. 1959. Searching for interstellar communications. Science 184: 844-846.

Cole, Lamont C. 1958. The Ecosphere. Sci. Amer. 198(4): 83-92.

Cousins, Norman. 1979. Anatomy of an Illness as Perceived by the Patient: Reflections on Healing and Regeneration. 173 p. W. W. Norten & Co: New York.

Darwin, Charles. (Originally published in 1859). 1963.The Origin of Species. With an introduction by W. R. Thompson. xxx + 488 p. Everyday's Library, Dutton: New York.

Dawkins, Richard. 1976. The Selfish Gene. xi + 224 p. Oxford University Press: New York and Oxford.

De Santas, Stefan. 1995. Nature and Man: The

Hindu Perspectives. (First English edition.) xxvi + 485 p., plus appendix xxix. Sociccus & Dilip Rumer Publ., Varanasi: India.

Douglas, William O. 1965. A Wilderness Bill of Rights. 192 p. Little, Brown and Company: Boston.

Drake, Frank and Dava Sabel. 1992. Is Anyone Out There? The Scientific Search for Extraterrestrial Intelligence. xv + 272 p. Delacorte Press: Bantam Group, Doubleday: New York.

Ellul, Jacques. 1965 (originally published in French in 1954). Propaganda: The Formation of Men's Attitudes. Translated from the French by Konrad Kellen and Jean Lerner. With an introduction by Konrad Kellen. xxii + 320 p. Vintage Books, Random House: New York.

Falk, Richard A. 1971. This Plundered Planet: Prospects and Proposals for Human Survival. x + 495 p. Random House: New York.

Freud, Sigmund. (Originally published in German in 1930.) 1961. Civilization and its Discontents. Newly translated from the German and edited by James Strachey. xiii + 109 p. W. W. Norton & Co: New York.

Goodess, C. M., Palutikof, J. P. and T. D. Davies. 1992. The Nature and Causes of Climate Change: Assessing the long-term future. xiii + 248 p. Lewis Publishers: Boca Raton; Belhaven Press: London.

Gore, Al. 2006. An Inconvenient Truth: The Planetary Emerging of Global Warming and What We Can Do About It. 327 p. Rodale, 23 E, Minor St., Emmaus, Pa. 18098.

Grinevald, Jacques. 2004. Consult the Internet for Grinevald, Université de Génève, Institut Universitaire d'études de Développement.

Halliday, James. 1948. Psychosocial Medicine: A Study of the Sick Society. 278 p. W. W. Norten & Co: New York.

Harlow, Harry F. 1958. The Nature of Love. Amer. Psychologist 13: 573-685. This article is also available on the Internet under the heading of Classics in the History of Psychology.

Hawken, Paul, Amory Lovins and L. Hunter Lovins. 1999. Natural Capitalism: Creating the Next Industrial Revolution. xix + 396 p. Little, Brown and Co: Boston.

Heilbronner, Robert L. 1967. The Worldly Philosophers. Third edition, newly revised. 320 p. A Clarion Book, Simon and Schuster: New York.

Henderson, L. J. (Originally published in 1913). 1958. The Fitness of the Environment: An Inquiry into the Biological Significance of the Properties of Matter. With an Introduction by George Wald. 317 p. Beacon Press: Boston.

Hoffer, Eric. 1969. The Temper of our Time. xi + 186 p. Harper & Row: New York.

Hutchinson, G. E. 1948. On Living in the Biosphere. Scientific Monthly. LXVII: 393-398.

Illich, Ivan. 1970. Deschooling Society. xx + 116 p. Harper & Rowe: New York.

Jacobs, Jane. 1984. Cities and the Growth of Nations: Principles of Common Life. ix + 257 p. Random House: New York.

James, William. (Originally published in 1890 by Henry Holt and Company.) 1950. The Principles of Psychology. Vol. I, vii + 689 p. Vol .II, vi + 688 p. Reprinted by Dover Publications: New York.

James, William. (Originally published in 1897). 1956. "The Sentiment of Rationality" p. 63-110 *in* "The Will to Believe and Other Essays in Popular Philosophy." xvii + 332 p. Dover Publications, Inc: New York.

James, William. (Originally published in 1902.) 1958. The Varieties of Religious Experience. With a Forward by Jacques Barzun. xviii + 406 p. A Mentor Book, The New American Library: New York.

James, William. 1912. Essays in Radical Empiricism. x + 283 p. Longmans, Green, and Co: New York.

Jeans, Sir James. 1930. The Mysterious Universe. ix + 154 p. Cambridge University Press: Cambridge.

Kant, Emmanuel. (Originally published in German

in 1795). 1972. Perpetual Peace. A Philosophical Essay. Translated with an introduction and notes by M. Campbell Smith. With a preface by Robert Latta. With a new Introduction for the Garland Edition by Samuel M. Thompson. xi + 203 p, Garland Publishing Inc: New York.

Lafreniere, G. F. 1983. Rousseau's first discourse and the idea of progress. Willamette J. Liberal Arts 1: 7-26.

Lafreniere, G. F. 1989. The redefinition of progress. Willamette J. Liberal Arts 4(2): 73-93.

Lamarck, J.-B. (Originally published in French in 1809 as Philosophie Zoologique.) 1965. Zoological Philosophy. Translated with an introduction by Hugh Elliot. xcii + 410 p. Hafner Publ. Co: New York.

Leiss, William. 1972. The Domination of Nature. xvii + 242 p. George Brazillier: New York.

Leopold, Aldo. (Originally published in 1949.) 1966. A Sand County Almanac. xix + 95 p. Ballantine Books: New York.

Livingston, John A. 1994. Rogue Primate: An explanation of human domestication. ix + 229 p. Kay Porter Books: Toronto.

Love, Glen A. 2003. Practical Ecocriticism: Literature, Biology, and the Environment. viii + 213 p. University of Virginia Press: Chalottesville & London.

Lovelock, James, 1991. Healing Gaia. 192 p. Harmony Books: New York.

Mata, Francisco J., Larry J. Onisto and J. R. Vallentyne. 1994. Consumption: The Other Side of Population for Development. Prepared by the Commission on Global Governance and the Earth Council for the International Conference on Population and Development, Cairo, 3-14 September, 1994. 17 p.

Maxwell, Mary. 1990. Morality Among Nations: An Evolutionary View. xi +198 p. State University of New York Press: Albany.

Mayr, Ernst and Carl Sagan. 1996. The Search for Extraterrestrial Intelligence: Scientific Quest or Hopeful Folly? The Planetary Report 16(3): 4-7, 11-13.

McFague, Sallie. 1993. The Body of God: An Ecological Theology. xii + 274 p. Fortress Press: Minneapolis.

McHarg, Ian. 1969. Design with Nature. viii + 198 p. Natural History Press: Garden City, New Jersey.

McNeish, Diana and Tony Newman. 2002. Involving young people in decision making. p186-204 *in* What Works for Children? Ed. by Diana McNeish, Tony Newman and Helen Roberts. Open University Press: Buckingham, Philadelphia.

Meadows, D. H., Meadows, D. L., Randers, J. and

W. W. Behrens, III. 1972. Limits to Growth: A Report for the Club of Rome's Project on The Predicament of Mankind. 207 p. A Potomac Associates Book, 2nd ed., revised. The New American Library: New York.

Metzner, Ralph. 1995. The psychopathology of the human-nature relationship. p. 55-67 in "Ecopsychology: Restoring the Earth, Healing the Mind" ed. by Roszak, T., N. E. Gomes and A. D. Kanner. Sierra Club Books: San Francisco.

Montagu, Ashley. 1971. Touching: The Human Significance of Skin. xv + 494 p. Perennial Library, Harper and Row Publishers: New York.

Mumford, Lewis. 1934. Technics and Civilization. 495 p. Harcourt, Brace and Company: New York.

Mumford, Lewis. 1938. The Culture of Cities. xii + 586 p. Harcourt, Brace & World, Inc: New York.

Mumford, Lewis. 1954. In the Name of Sanity. 244 p. Harcourt, Brace and Company: New York.

Petersen, A. F. 1983. On downward causation in biological and behavioral systems. Publ. Staz. Zool. Napoli 5(1): 69-86.

Platt, John. 1969. What we must do. Science 166: 1115-1121.

Postman, Neil. 1992. Technopoly: The Surrender of

Culture to Technology. xii + 222 p. Alfred A. Knopf: New York.

Postman, Neil. 1996. The End of Education: redefining the nature of school. xi + 209 p. Alfred A. Knopf: New York.

Ray, Paul H. and Sherry Ruth Anderson. 2000. The Cultural Creatives: How 50 million people are changing the world. xviii + 370 p. Harmony Books: New York.

Rechnitzer, Peter. 1994. R. M. Bucke: Journey to Cosmic Consciousness. 256 p. Associated Medical Services Ltd. and Fitzhenry & Whiteside: Markham, Ontario.

Rowe, Stan. 1992. Biological Fallacy: Life Equals Organisms. BioScience 42(6): 394.

Shepard, Paul. 1983. Nature and Madness. xii + 178 p. Sierra Club Books: San Francisco.

Smith, Adam. (Originally published in 1776.) 1970. The Wealth of Nations. Books I-III. A reprint with an introduction by Adrian Skinner. 535 p. Penguin Books: New York.

Smuts, J. C. 1927. Holism and Evolution. xiii + 368 p. Macmillan and Co: London.

Stone, Christopher D. 1996. Should Trees Have Standing? Twenty-fifth Anniversary Edition with other essays on law, morals and protection of the environment. 181 p. Oceana Publications Inc: Dubbs Ferry, N.Y.

Sun Tzu. (Originally from China about 300 B.C.)

1963. The Art of War. Translated and with an introduction by Samuel B. Griffith and with a forward by Liddell B. Hart. xvii + 197 p. Oxford University Press: Oxford.

Suzuki, David and Holly Dressel. 2002. Good News for a Change: Hope for a Troubled Planet. 0 + 398 p. Stoddart Publ. Co: Toronto.

Tansley, A.G. 1935. The use and abuse of vegetational concepts and terms. Ecology 16: 284-307.

Teilhard de Chardin, Pierre. (Originally published in French in 1956.) 1959. The Phenomenon of Man. Translated by Bernard Wall. 320 p. Harper & Row: New York.

Turley, Peter T. 1977. Peirce's Cosmology. 126 p. Philosophical Library: New York.

Vallentyne, J. R. 1978, Today is Yesterday's Tomorrow. Verh. Internat. Verein. Limnol. 20: 1-12.

Vallentyne, J. R., Strickler, J. R. and Nicholas Polunin. 1980. Proposal: International Year of the Biosphere. Env. Conserv. 7(1): 2.

Vallentyne, J. R. and H. L. Tracy. 1972. A new term introduced at First Conference on the Environmental Future. Biol. Conserv. 4 (5): 372. (Reprinted in Bull. Atomic Scientists: 29(5): 24; 1973).

Vernadsky, V. I. 1926. Biosfera. 150 p. Nauchnoe Khimikoteknicheskoe Izdatelstvo: Leningrad (St. Petersberg).

Wagar, W. W. 1961. H.G. Wells and the World State. x + 301 p. Books for Libraries Press: Freeport, New York.

Wells, H.G. 1928. The Open Conspiracy. Blueprints for a World Revolution. 156 p. Victor Gollancz Ltd: London.

Whitehead, Alfred North. (Originally published in 1929.) 1967. The Aims of Education and other essays. vii + 165 p. A Free Press Paperback. Collier-Macmillan Canada: Toronto.

Wood, Harold W., Jr. 1985. Modern pantheism as an approach to environmental ethics. Environmental Ethics 7: 151-162. Also, search the Internet for pantheism.

Wooding, G. Scott. 2005. The Parenting Crisis. viii + 224 p. Fitzhenry and Whiteside: Markham, Ontario, and Allston, Mass.

INDEX TO TEXT AND NOTES

THE SORCERER LURKS WITHIN.

On July 9 1968 John B. Calhoun, American behavioral ecologist, introduced eight mice into a technologically designed walled enclosure that fulfilled all the wants and needs of mice except migration in and out. Over a 4 ½ year period the population exploded into a colony of 2200 mice and then slowly and inexorably declined to extinction. Deprived of motherly love early in life, and denied access to social roles later in life, young mice grew up without knowing how to behave as mice. This book examines whether a similar fate could be in store for Human Utopia.

Part I compares the suicidal effects of runaway growth in Mouse Utopia to a similar sequence of events in Human Utopia.

Part II describes our dependence on the Biosphere.

Part III shows how the runaway growth of technology and population is creating havoc in our species and the Biosphere. How an inner life-force symbolized as "the Sorcerer" causes populations to self-destruct by satisfying internal desires beyond their useful times. As the Sorcerer's apprentices, we have been preparing the ground for a deluge of crises beyond human control.

Part IV describes how the Sorcerer works and how to control his destructive traits. Political, spiritual and behavioral opportunities are identified that could be overlooked, misinterpreted or ignored in steering a course toward coevolution of humanity and the Biosphere.

ISBN 1-41205633-0

9 781412 056335

TRAFFORD

Trafford Publishing
Suite 6E-2333 Government St.,
Victoria, BC Canada V8T 4P4
www.trafford.com

Cover design by: Art Department Design
www.artdepartmentdesign.com